Colloid and Interface Chemistry for Water Quality Control

Colloid and Interface Chemistry for Water Quality Control

Qing Chang
Lanzhou Jiaotong University
Lanzhou, Gansu, China

ELSEVIER

Chemical Industry Press

AMSTERDAM • BOSTON • HEIDELBERG • LONDON
NEW YORK • OXFORD • PARIS • SAN DIEGO
SAN FRANCISCO • SINGAPORE • SYDNEY • TOKYO
Academic Press is an imprint of Elsevier

AP

Academic Press is an imprint of Elsevier
125 London Wall, London EC2Y 5AS, UK
525 B Street, Suite 1800, San Diego, CA 92101-4495, USA
50 Hampshire Street, 5th Floor, Cambridge, MA 02139, USA
The Boulevard, Langford Lane, Kidlington, Oxford OX5 1GB, UK

British Library Cataloguing-in-Publication Data
A catalogue record for this book is available from the British Library

Library of Congress Cataloging-in-Publication Data
A catalog record for this book is available from the Library of Congress

ISBN: 978-0-12-809315-3

For information on all Academic Press publications
visit our website at https://www.elsevier.com/

Publisher: John Fedor
Acquisition Editor: Simon Tian
Editorial Project Manager: Naomi Robertson
Production Project Manager: Julie-Ann Stansfield
Designer: Victoria Pearson Esser

Typeset by Thomson Digital

Contents

Preface

Colloid and interface chemistry is one of the most important scientific bases of water quality control. It is related to nearly all of the principles of water treatment, including coagulation-flocculation, sedimentation, adsorption, filtration, flotation, oil separation, membrane separation, sludge dewatering, scaling-corrosion inhibitor, and environmental catalysis. The need for water quality control has quickly increased for decades because people pay more and more attention to drinking water safety and water environment protection. Decision-making in matters relating to water quality control requires a sound understanding of basic principles of colloid and interface chemistry; therefore, the publication of this book is quite relevant at present.

This book was completed on the basis of my teaching materials in the School of Environmental and Municipal Engineering at Lanzhou Jiaotong University of China. It provides students the most basic knowledge of colloid and interface chemistry. Chapters 1–7 introduce colloid chemistry including Chapter 1 "Brief History of Colloid and Interface Chemistry and Basic Concepts," Chapter 2 "Diffusion and Brown Motion," Chapter 3 "Sedimentation," Chapter 4 "Osmotic Pressure," Chapter 5 "Optical Properties," Chapter 6 "Rheology Properties," Chapter 7 "Electric Properties"; Chapters 8–10 introduce interface chemistry including Chapter 8 "Surface of Liquid," Chapter 9 "Surface of Solution," and Chapter 10 "Surface of Solids"; Chapter 11 "Emulsion, Foam, and Gel" introduces the coarse dispersion system. As a classical science, colloid and interface chemistry has been developing for more than a century and encompasses a large amount of knowledge, but this book focuses on that which can be applied to water treatment and is closely related to the practices of water treatment and water purification; therefore, students and specialists in the area will find it very useful.

This book is characterized by concision, and the main basic principles and theories are introduced and described. It is easy to read and understand. I think textbooks should not be designed simply as a collection of facts but rather an introduction to ways of thinking about the world. Questions and exercises as well as their solutions have been given throughout the book so that readers can check their understanding of the text. Both traditional and some newly developed knowledge are included to reflect the latest advances of the discipline.

I hope this book will be used as a reference book or a text book for undergraduate students and graduate students who are majoring in water quality control to help them understand the problems and difficult points in their studies. It could also be used as a reference for researchers and engineers as well as graduate students in other majors for further study. In addition, it is quite suitable for researchers and engineers to quickly consult relevant information that they encounter in their work because of the book's brevity and clarity.

It is always a pleasure to thank my editor Gang Wu who has contributed so generously with illustrations and helpful suggestions to improve the book. My appreciation also goes to Dr Pengyu Liu for his work on the figures. Finally, I am very grateful for the financial support from the National Natural Science Foundation of China (No. 21277065) in preparation of this book.

Qing Chang
Lanzhou Jiaotong University
Lanzhou, Gansu, China
October, 2015

Chapter 1

Brief History of Colloid and Interface Chemistry and Basic Concepts

Chapter Outline

1.1 ORIGIN OF THE TERM "COLLOID"

The system formed by dispersing one kind of substance in another kind of substance is referred to as a dispersion system. The former is called the dispersed phase, and the latter is the dispersion medium. For example, dispersing sodium chloride (NaCl) in water is a dispersion system. Here the sodium chloride is the dispersed phase, and the water is the dispersion medium. Generally, the dispersed phase contains particles of various sizes. The smaller the particle, the higher the dispersion degree; the larger the particle, the lower the dispersion degree.

As early as 1861, Graham studied the permeability of various dispersion systems through parchment paper. The device for this study is shown in Fig. 1.1.

According to the permeability characteristics obtained in Graham's experiment, the dispersion systems studied were divided into two types as follows:

1. *True solution*: The diffusion is faster, and the permeation is easier. Crystal is obtained after evaporation to dryness, for example, sugar, inorganic salt, and urea.
2. *Colloid*: The diffusion is slower, and the permeation is more difficult. An amorphous substance is obtained after evaporation to dryness. "Colloid' means the substance does not easily permeate through parchment paper, for example, gelatin, aluminum hydroxide, and polysilicic acid.

Colloid and Interface Chemistry for Water Quality Control. http://dx.doi.org/10.1016/B978-0-12-809315-3.00001-3

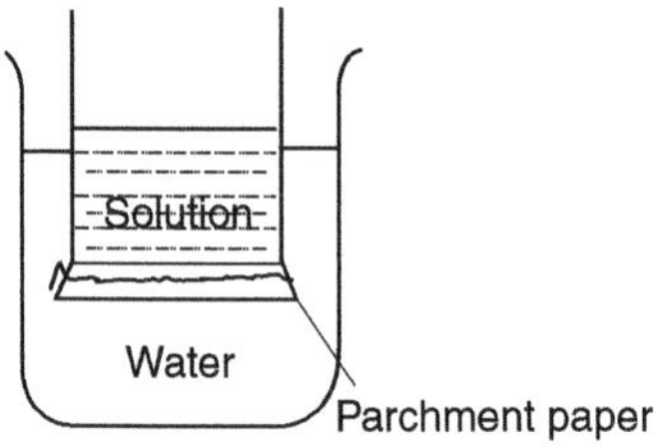

FIGURE 1.1 Device for the Graham experiment.

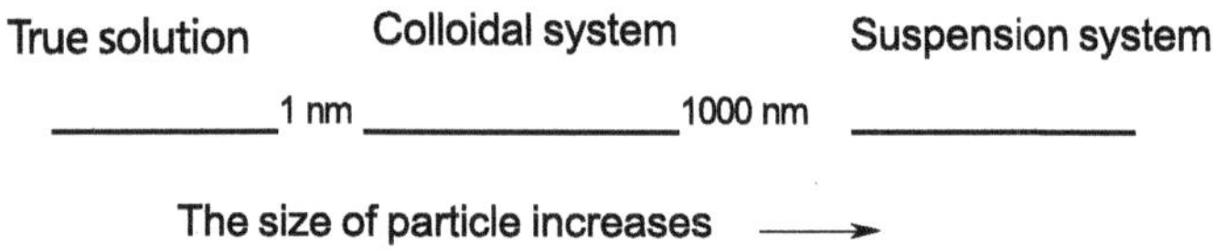

FIGURE 1.2 A type of dispersion system.

Afterward, the experiment of Beŭmph proved that the two types of dispersion systems can be transformed into each other; for example, if we disperse NaCl in water, we will obtain a true solution; if we disperse NaCl in alcohol, we will obtain a colloid system. There are many such instances that indicate that the terms "colloid" and "true solution" do not represent the essence of the material; instead, they represent two different forms of the substance. The difference between them is only the size of the particles or the dispersion degree, as shown in Fig. 1.2.

Now it is widely accepted that the system in which a particle has at least one dimension in the range of 1–1000 nm is considered a colloid system. Therefore, the size of the particle is the only criterion used to determine if the dispersion system is a colloid system. Linear dimension rather than particle weight, or the number of atoms in a particle, defines the colloid size range.

1.2 CLASSIFICATION OF THE COLLOIDAL SYSTEM

Although the size of particles in a colloid system is in the range of 1–1000 nm, the properties are not always the same. According to the different properties of particles, the colloid system can be grouped into two general classifications:

1. *Sol*: There are interface and interface-free energies between the dispersed phase and the dispersion medium. It is a thermodynamically unstable, irreversible system and not easily reconstituted after phase separation. This is often called a lyophobic colloid.

2. *The solution of macromolecular material*: There are no interface and interface-free energies between the dispersed phase and the dispersion medium. It is a thermodynamically stable system and can be recovered after separation. This is often called a lyophilic colloid.

There are some similar properties and some different properties between the two kinds of colloid systems. The properties that are related to particle size are similar, but the properties that are related to interface are different. For example, dynamics properties, optical properties, rheology properties, etc., are similar, but electrical properties, adsorption properties, etc., are different.

Although the size of the particles in the suspension system is greater than that in the colloid system, there is also greater interface between the dispersed phase and the dispersion medium in the suspension system, and many properties of the suspension system are similar to that of the colloid system; therefore, the study on the suspension system is generally included in the category of colloid chemistry.

Generally speaking, natural water, industrial west water, and domestic sewage not only contain dissolved salts, becoming true solutions, but also contain colloid and suspension particles, becoming colloid systems and suspension systems; therefore, they are complicated and integrated systems. Because particle separation is one of the most important tasks of water treatment, colloid and interface chemistry is very important for researchers and engineers who engage in water treatment and water environment protection.

In another way, colloid systems can also be grouped according to the kinds of dispersed phases and dispersion mediums, as in Table 1.1.

TABLE 1.1 Classification of Colloid Systems

Dispersed phase	Dispersion medium	Name	Example
Liquid	Gas	Liquid aerosol	Fog
Solid	Gas	Solid aerosol	Smoke, dust
Gas	Liquid	Foam	Foam on soap solutions
Liquid	Liquid	Emulsion	Milk
Solid	Liquid	Sol	Paint
Gas	Solid	Solid foam	Foamed plastic
Liquid	Solid	Solid emulsion, gel	Pearl
Solid	Solid	Solid suspension system	Alloy

As far as water treatment is concerned, sol, emulsion, and foam are very important. The natural turbid water belongs to sol and suspension dispersion systems. The industrial or domestic oil wastewater often forms emulsion. The wastewater that contains protein or surfactant generally produces foam. Essentially, the effect of water treatment is to destroy the stabilities of the colloidal dispersion systems.

1.3 DISPERSION DEGREE AND SPECIFIC SURFACE AREA

Generally, the dispersion degree of multiphase dispersion systems can be represented by the specific surface area A_0:

$$A_0 = \frac{A}{V} \tag{1.1}$$

where A is the total surface area of dispersed substance, and V is the total volume of dispersed substance. Therefore, the specific surface area is the surface area of the substance's unit volume (or unit mass). Its value quickly increases with the decrease in particle size. When a cube 1 cm in length is divided into smaller cubes, the specific surface area will increase as shown in Table 1.2.

It can be seen that the smaller the cubes, the greater the specific surface area. As we know, the size of colloidal particles is within the range of 1–1000 nm; therefore, the colloid system has great specific surface area, and its surface effect is very significant. In fact, many principles of colloid chemistry belong to interface chemistry.

TABLE 1.2 Division of Cubes and Increase of Specific Surface Areas

Length of side (cm)	Cube numbers	Specific surface area (cm^{-1})
1	1	6
10^{-1}	10^3	6×10
10^{-2}	10^6	6×10^2
10^{-3}	10^9	6×10^3
10^{-4}	10^{12}	6×10^4
10^{-5}	10^{15}	6×10^5
10^{-6}	10^{18}	6×10^6
10^{-7}	10^{21}	6×10^7

Example 1.1 Variation of Specific Surface Area with Geometry

A material of density ρ exists as uniform cylindrical particles of radius R_c and length L. Derive an expression for A_0 for this material and examine the limiting forms when either R_c or L is very small.

Solution: The area of each cylindrical particle equals the sum of the areas of both ends and the cylindrical surface:

$$A = 2\left(\pi R_c^2\right) + 2\pi R_c L$$

The volume of each cylindrical particle equals $\pi R_c^2 L$, and its mass is given by $\rho \pi R_c^2 L$.

For an array of n cylindrical particles, the total area per total mass equals A_0 and is given by

$$\begin{aligned} A_0 &= \frac{n\left(2\pi R_c^2 + 2\pi R_c L\right)}{n\rho\pi R_c^2 L} \\ &= \frac{2\left(R_c^2 + R_c L\right)}{\rho R_c^2 L} \\ &= \left(\frac{2}{\rho}\right)\left(\frac{1}{R_c} + \frac{1}{L}\right) \end{aligned}$$

For a thin rod, $L \gg R_c$:

$$A_0 \approx \frac{2}{\rho}\frac{1}{R_c}$$

For a flat disk, $R_c \gg L$:

$$A_0 \approx \frac{2}{\rho}\frac{1}{L}$$

1.4 SHAPE OF COLLOIDAL PARTICLES

Particle shape has very important effects on the properties of colloid systems. For example, polystyrene latex has very low viscosity and flows easily, even if its concentration increases to 10~20%, because the particle is spherical; vanadic oxide has very high viscosity and cannot flow, even if its concentration decreases to 0.01%, because the particle is filamentous.

Polyacrylamide is usually used as the flocculant in water treatment. Its macromolecule is a very long chain, and its viscosity is very high. Before using, it should be diluted in order to add it into water easily; for example, the concentration should be in the range of 0.05~0.1% in laboratory experiments.

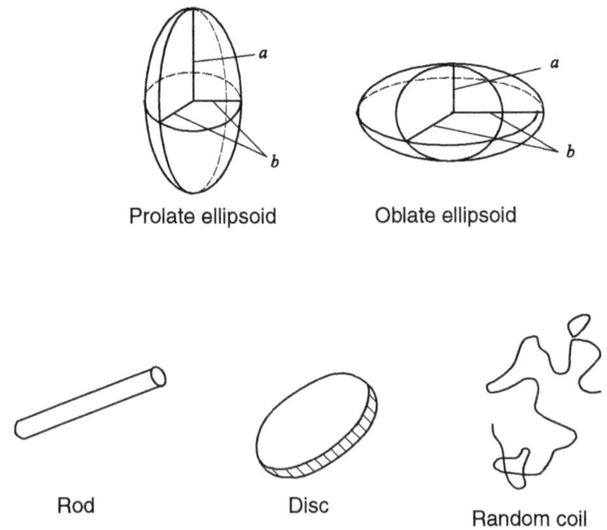

FIGURE 1.3 Representations of nonspherical particles.

The particle shape can be grouped into three types as follows:

1. *Sphere*: It is generally characterized by its radius.
2. *Ellipsoids of revolution*: They are generally characterized by their axial ratio, which is the ratio of the single half-axis (a) to the radius of revolution (b), as shown in Fig. 1.3. The axial ratio is greater than 1 for a prolate ellipsoid and less than 1 for an oblate ellipsoid. If $a \gg b$, the particle will be a rod, and if $a \ll b$, the particle will be a disc.
3. *Threadlike macromolecule*: It is a random coil and can be characterized by its molecular weight.

1.5 POLYDISPERSITY AND THE AVERAGE SIZE OF COLLOIDAL PARTICLES

The dispersion system in which the molecular weights or the sizes of all particles are the same is termed a monodispersed system. This kind of system is very rare in nature, but synthetic polymers can possibly be this kind of system; for example, polystyrene latex might be a monodispersed system. The dispersion system in which the molecular weights or sizes of particles are not identical is termed polydispersed system. This kind of dispersion system is very general; for example, all natural waters and polluted natural waters are polydispersed. For polydispersed systems, the size of the particle is generally characterized by average diameter or average molecular weight. The former is suitable for sol, and the latter is suitable for macromolecule solutions.

1. Average diameter

There are three kinds of average diameters.

a. Number average diameter $\bar{d}_n$

Number average diameter is obtained by averaging the diameters to the particle numbers:

$$\bar{d}_n = \frac{\sum n_i d_i}{\sum n_i} = \sum f_i d_i \tag{1.2}$$

where n is the particle number, subscript i indicates the particle size, d is the particle diameter, and f is the weighting factor:

$$f_i = \frac{n_i}{\sum n_i} \tag{1.3}$$

The average diameter obtained using a microscope is the number average diameter.

b. Area average diameter $\bar{d}_s$

Area average diameter is a root mean square value converted from the average particle areas obtained from the adsorption experiment. If A represents the area of the particle, the average area of particles can be expressed as

$$\bar{A} = \frac{\sum n_i A_i}{\sum n_i} = \frac{\sum n_i \pi d_i^2}{\sum n_i} = \pi \sum f_i d_i^2 \tag{1.4}$$

or

$$\bar{A} = \pi \left(\bar{d}_s\right)^2 \tag{1.5}$$

Combining Eqs. (1.4) and (1.5) gives

$$\pi \sum f_i d_i^2 = \pi \left(\bar{d}_s\right)^2$$

$$\bar{d}_s = \left(\sum f_i d_i^2\right)^{\frac{1}{2}} \tag{1.6}$$

c. Volume average diameter $\bar{d}_v$

Volume average diameter is converted from the average particle volume obtained from the density. If V represents the volume of particles, the average volume of particles can be expressed as

$$\bar{V} = \frac{\sum n_i V_i}{\sum n_i} = \frac{\sum n_i \frac{\pi}{6} d_i^3}{\sum n_i} = \frac{\pi}{6} \sum f_i d_i^3 \tag{1.7}$$

or

$$\bar{V} = \frac{\pi}{6}\left(\bar{d}_v\right)^3 \tag{1.8}$$

Combining Eqs. (1.5) and (1.6) gives

$$\frac{\pi}{6}\Sigma f_i d_i^3 = \frac{\pi}{6}\left(\bar{d}_v\right)^3$$

$$\bar{d}_v = \left(\Sigma f_i d_i^3\right)^{\frac{1}{3}} \tag{1.9}$$

For the polydispersion system, we have

$$\bar{d}_n < \bar{d}_s < \bar{d}_v \tag{1.10}$$

For the monodispersion system, we have

$$\bar{d}_n = \bar{d}_s = \bar{d}_v \tag{1.11}$$

The ratio $\bar{d}_s / \bar{d}_n$ is used to show the significance of the polydispersity of the system. The more $\bar{d}_s / \bar{d}_n$ deviates from 1.0, the more significant the dispersity of the system.

2. Average molecular weight

There are three kinds of average molecular weights

a. Number average molecular weight $\overline{M_n}$

Number average molecular weight is obtained by averaging the particle weights to the particle numbers:

$$\overline{M_n} = \frac{W}{\Sigma n_i} = \frac{\Sigma n_i M_i}{\Sigma n_i} = \Sigma f_i M_i \tag{1.12}$$

where n is the particle number, subscript i indicates the particle size, W is the total weight of the particles, M is the molecular weight of the particle, and f is the weighting factor:

$$f_i = \frac{n_i}{\Sigma n_i} \tag{1.13}$$

Generally speaking, the numbers of smaller particles are greater, leading to greater weighting factors; therefore, smaller particles make greater contributions to the number average molecular weight. As a result, the value of the number average molecular weight is closer to the molecular weights of smaller particles. The molecular weight obtained from the colligative property measurement, for example, osmotic pressure, boiling point rising, and freezing point depression, is the number average molecular weight.

b. Weight average molecular weight $\overline{M_w}$

Weight average molecular weight is obtained by averaging the molecular weights to the particle weights:

$$\overline{M_w} = \frac{\Sigma W_i M_i}{\Sigma W_i} = \Sigma f_i' M_i \tag{1.14}$$

where W is the weight of the particle and f_i' is the weighting factor:

$$f_i' = \frac{W_i}{\Sigma W_i} \tag{1.15}$$

Generally speaking, the weights of larger particles are greater, leading to greater weighting factors; therefore, larger particles make greater contributions to weight average molecular weight. As a result, the value of weight average molecular weight is closer to the molecular weights of larger particles. In addition, weight average molecular weight can be calculated as follows:

$$\overline{M_w} = \frac{\Sigma W_i M_i}{\Sigma W_i} = \frac{\Sigma n_i M_i^2}{\Sigma n_i M_i} = \frac{\Sigma f_i M_i^2}{\Sigma f_i M_i} \tag{1.16}$$

The molecular weight obtained from light-scattering measurements is the weight average molecular weight.

c. Z average molecular weight $\overline{M_z}$

Z average molecular weight is obtained by averaging the molecular weights to the product of the particle weights and particle molecular weights:

$$\overline{M_z} = \frac{\Sigma W_i M_i^2}{\Sigma W_i M_i} = \Sigma f_i'' M_i \tag{1.17}$$

where f_i'' is the weighting factor:

$$f_i'' = \frac{W_i M_i}{\Sigma W_i M_i} \tag{1.18}$$

Generally speaking, the product of weights and molecular weights of larger particles are even greater, leading to even greater weighting factors; therefore, larger particles make even greater contributions to Z average molecular weight. As a result, the value of Z average molecular weight is even closer to the molecular weights of larger particles. In addition, Z average molecular weight can be calculated as follows:

$$\overline{M_z} = \frac{\Sigma W_i M_i^2}{\Sigma W_i M_i} = \frac{\Sigma n_i M_i M_i^2}{\Sigma n_i M_i M_i} = \frac{\Sigma f_i M_i^3}{\Sigma f_i M_i^2} \tag{1.19}$$

The molecular weight obtained from sedimentation equilibrium measurement is Z average molecular weight.

Comparing the aforementioned three kinds of molecular weights for the polydispersion system, we have

$$\overline{M_z} > \overline{M_w} > \overline{M_n} \tag{1.20}$$

Comparing the aforementioned three kinds of molecular weights for the monodispersion system, we have

$$\overline{M_z} = \overline{M_w} = \overline{M_n}$$

The ratio $\overline{M_w}/\overline{M_n}$ is used to show the significance of the polydispersity of the system. The more $\overline{M_w}/\overline{M_n}$ deviates from 1.0, the more significant the dispersity of the system.

Example 1.2 Molecular Weight of a Polymer

There are two fractions for a polymer, one has the molecular weight of 1×10^3, and the other has the molecular weight of 1×10^5. Take 1 g of polymers from these two fractions respectively and mix them. Calculate the number average molecular weight, the weight average molecular weight, and the Z average molecular weight of this mixture. What conclusion can be obtained from the calculated results?

Solution 1. $f_1 = \dfrac{\dfrac{1}{1\times 10^3}}{\dfrac{1}{1\times 10^3} + \dfrac{1}{1\times 10^5}} \approx 1$

$$f_2 = \frac{\dfrac{1}{1\times 10^5}}{\dfrac{1}{1\times 10^3} + \dfrac{1}{1\times 10^5}} \approx 1\times 10^{-2}$$

$$\overline{M_n} = \Sigma f_i M_i \approx 1\times 10^3 + 1\times 10^{-2} \times 10^5 \approx 2\times 10^3$$

2. $f_1' = \dfrac{1}{1+1} = \dfrac{1}{2}$

$$f_1' = \frac{1}{1+1} = \frac{1}{2}$$

$$\overline{M_w} = \Sigma f_i' M_i = \frac{1}{2}\times 1\times 10^3 + \frac{1}{2}\times 1\times 10^5 = 5\times 10^4$$

$$3.\ f_1'' = \frac{1\times1\times10^3}{1\times1\times10^3+1\times1\times10^5} \approx 10^{-2}$$

$$f_2'' = \frac{1\times1\times10^5}{1\times1\times10^3+1\times1\times10^5} \approx 1$$

$$\overline{M_z} = \Sigma f_i'' \approx 10^{-2}\times1\times10^3+1\times1\times10^5 \approx 10^5$$

Conclusions

The value of the number average molecular weight is closer to the molecular weight of the fraction with lower molecular weight; the value of weight average molecular weight is closer to the molecular weights of the fraction with higher molecular weight; and the value of Z average molecular weight is even closer to the molecular weight of the fraction with higher molecular weight. In addition, comparing the aforementioned three kinds of molecular weights, we have an order as follows:

$$\overline{M_z} > \overline{M_w} > \overline{M_n}$$

1.6 COLLOIDAL POLLUTANTS IN NATURAL WATERS

There are many kinds of colloidal particles in natural waters, for example, various ores; hydrous metal oxides; hydrous silicon oxides; humic substances; proteins; oil drops; air bubbles; surfactant hemicolloids; and biocolloids, including alga, bacterium, and virus.

1. *Ore particle*: The common colloidal ore particles in natural waters are silicate ores, including quartz, feldspar, mica, and clay minerals that include montmorillonite, illite, and kaolin. Quartz and feldspar are not easy to break, and their sizes are greater. Mica, montmorillonite, and kaolin are easy to break, and their sizes are tiny. The clay minerals in natural waters are the most significant properties of colloid chemistry; they are secondary minerals formed from primary minerals during the weathering process, mainly aluminum and magnesium silicates, and are of laminar structure.
2. *Hydrous metal oxide*: Common colloidal hydrous metal oxide particles in natural waters are hydrous ferric oxide, hydrous aluminum oxide, hydrous manganese oxide, hydrous silicon oxide (metalloid), etc. They are commonly considered to be the polynuclear complexes that are the intermediate products during the transformation process of aquometal from ions into hydroxide. Many multivalent hydrous oxides are amphoteric due to the acid–base equilibrium involved in the hydrolysis reactions of aquometal ions. Thus hydrogen ions and hydroxide ions are the primary potential-determining ions for hydrous oxides; therefore, the charge of the hydrous oxide depends on the pH value of the medium.

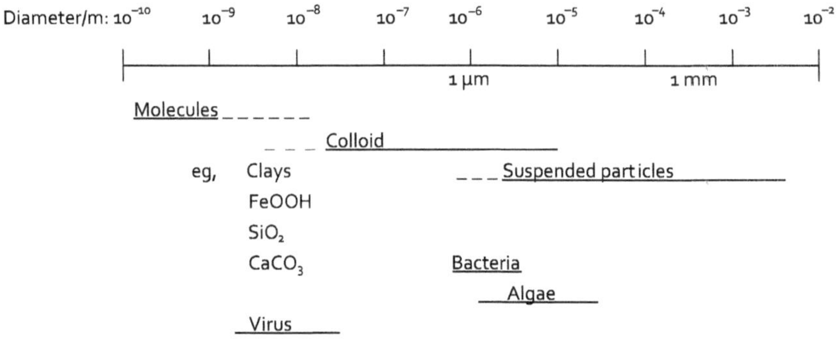

FIGURE 1.4 Size spectrum of waterborne particles.

3. *Humic substances*: Humic substances are considered to be the most important part of natural organic substances in natural waters. Various biological materials can be transformed to form the humic substances. They are polymers of phenol and quinine with hydroxyl and carboxyl. The molecular weight of humic substances is in the range of 300 ~ 30,000. Humic substances can be grouped into three kinds according to their solubility in acid and base. The constituent that is soluble in alkaline solution is humic acid; the constituent that is soluble not only in alkaline solution but also in acid solution is fulvic acid, and the constituent that is not soluble whether in alkaline solution or in acid solution is humin. Fulvic acid has lower molecular weight and more hydrophilic functional groups than humic acid and humin.

Fig. 1.4 shows the size spectrum of particles in natural waters. Generally speaking, the molecules are smaller than 10^{-3} μm; therefore, they cannot be removed from water by settling or filtration. Colloidal particles can pass through the pores of most filtration media, and their settling rates are very slow; therefore, they cannot be removed from water by either settling or filtration. Since the size of suspended particles is larger than 1 μm, they can be removed from water by settling or filtration.

Chapter 2

Diffusion and Brownian Motion

Chapter Outline

2.1 DIFFUSION

Diffusion is a spontaneous process by which a substance migrates from a region of high concentration to a region of lower concentration. The last result of diffusion is the uniform distribution of substance. Diffusion is induced because the chemical potential of substance in the region of high concentration is always higher than that in the region of lower concentration. The substance migrates from a region of high chemical potential to a region of lower chemical potential, which is referred to as chemical potential criteria. There are two laws for diffusion: Fick's first law and Fick's second law.

2.1.1 Fick's First Law

Let us define the diffusion direction as x, the mass of substance as m, the area of the cross plane perpendicular to the x-direction as A, the concentration of substance as c, the time as t and the proportionality factor as D. Then the mass of substance diffusing in the x-direction across an area A in a time dt is proportional to the concentration gradient at the plane in question dc/dx as shown in Eq. (2.1):

$$dm = -DA\frac{dc}{dx}dt \tag{2.1}$$

Since the diffusion direction is from the region of higher concentration to the region of lower concentration, dc/dx is always negative in the diffusion direction.

Colloid and Interface Chemistry for Water Quality Control. http://dx.doi.org/10.1016/B978-0-12-809315-3.00002-5

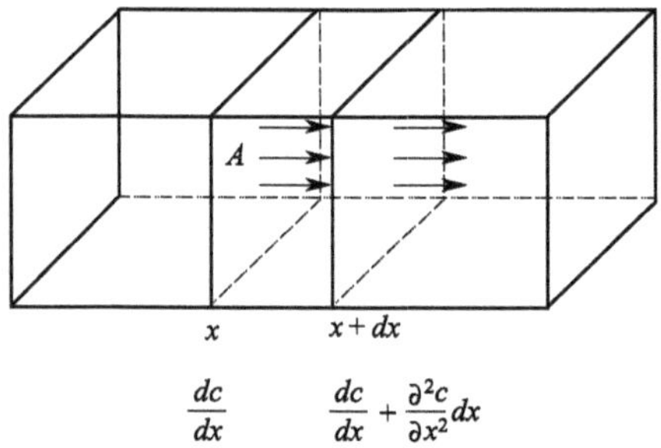

FIGURE 2.1 **Migration of substance in diffusion direction.**

Therefore, a minus sign is added in the equation. The proportionality factor D is termed diffusion coefficient, which can be simply thought of as the mass of substance diffusing across unit area at unit concentration gradient and in unit time.

2.1.2 Fick's Second Law

In a diffusion process, the concentration of substance in the region of higher concentration gradually decreases, and the concentration of substance in the region of lower concentration gradually increases. The uniform distribution of substance will reach a point where the concentrations of substance in all of regions will be equal at the end. Fick's second law describes the concentration change with time in this diffusion process.

Consider a cube in the diffusion direction as shown in Fig. 2.1, where x is the distance in diffusion direction, A the area of cross plane perpendicular to the x-direction, and m is the mass of substance. Here, let us study the change of the mass of substance in a small volume element Adx. Suppose dm of substance comes into the small volume element and dm' of substance leaves the small volume element in dt time; from Fick's first law, we have Eq. (2.1) as follows:

$$dm = -DA\frac{dc}{dx}dt$$

$$dm' = -DA\left(\frac{dc}{dx} + \frac{\partial^2 c}{\partial x^2}dx\right)dt \tag{2.2}$$

Where $\left(\frac{\partial^2 c}{\partial x^2}dx\right)$ is the micro change of concentration gradient from x to $x + dx$, therefore $\left(\frac{dc}{dx} + \frac{\partial^2 c}{\partial x^2}dx\right)$ is the concentration gradient at $x + dx$, and the increase of substance in the small volume element in dt time is obtained by Eq. (2.1) minus Eq. (2.2), that is,

$$dm - dm' = DA\left(\frac{\partial^2 c}{\partial x^2}dx\right)dt$$

thus

$$\frac{dm - dm'}{Adx} = D\frac{\partial^2 c}{\partial x^2}dt$$

Because Adx is the volume of the small volume element, we have

$$\frac{\partial c}{\partial t} = D\frac{\partial^2 c}{\partial x^2} \tag{2.3}$$

Eq. (2.3) is the expression of Fick's second law which shows the change rate of the concentration of substance in the small volume element with time. It is related to the second derivative of concentration to the distance in the diffusion direction and can be solved by integration.

Example 2.1 Diffusion

In following various figures, solid lines represent the initial distribution of concentration along distance, while dotted lines represent the distribution of concentration along distance after a short time of diffusion. Answer which of them is correct and what is the reason.

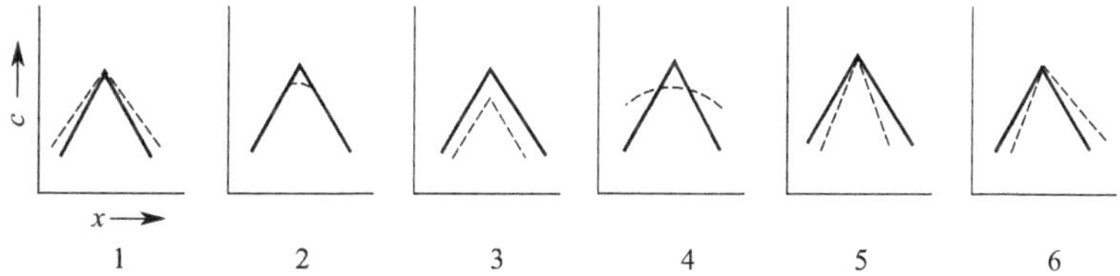

Solution

In the diffusion process, the concentration of substance in the region of higher concentration gradually decreases, the concentration of substance in the region of lower concentration gradually increases, and the uniform distribution of substance will reach at last, therefore the number 4 is correct.

2.2 BROWNIAN MOTION

In 1826, British botanist Brown placed some of pollen grains on the surface of water, and observed the suspended pollen grains making constant random motion. This random motion is referred to as Brownian motion. The explanation is given here as follows.

In essence, Brownian motion is the result of random collisions between suspended particles and the molecules of dispersion medium. For a suspended object that is large enough, the numbers of collisions from different directions are tremendous, therefore, can be considered to be nearly equal, hence, the acting forces from different directions nearly completely cancel, and the composition force nearly equals zero. However, for the suspended particles as small as colloid scale, the numbers of collisions from different directions are limited, therefore, cannot be considered to be nearly equal, therefore, the acting forces cannot cancel each other completely, the composition force do not equals zero, thus resulting the random movement of the particles.

Einstein's equation of Brownian motion is obtained as follows.

Firstly, let us introduce the concept of mean displacement. Although the movement of particle in Brownian motion is random, the displacement can be found after some time of observation. Define the projection of the displacement in horizontal ordinate direction is x, the mean displacement in the time t is

$$\overline{\Delta} = \left(\overline{x^2}\right)^{\frac{1}{2}} = \left(\sum_{i=1}^{n} \frac{x_i^2}{n}\right)^{\frac{1}{2}} \tag{2.4}$$

where i indicates the different particles, n is the number of particles. It can be seen that the mean displacement is a mean square root of displacement, and a constantly positive value. Suppose there is a cross plane AB in the tube which is full of colloidal dispersion system, and its area is S, as shown in Fig. 2.2. Again there are two liquid layers in the left and right side of plane AB respectively, and the thickness of each of the liquid layers equals the mean displacement of particles in time t, that is, $\overline{\Delta}$. In addition, the mean concentration of particle in the left layer is c_1, and the mean concentration of particle in the right layer is c_2, and $c_1 > c_2$.

Because the thickness of the liquid layer equals the mean displacement of particles in time t, even located in the left border of the left liquid layer the particle can reaches the cross plane AB in time t, that is, all of the particle in the

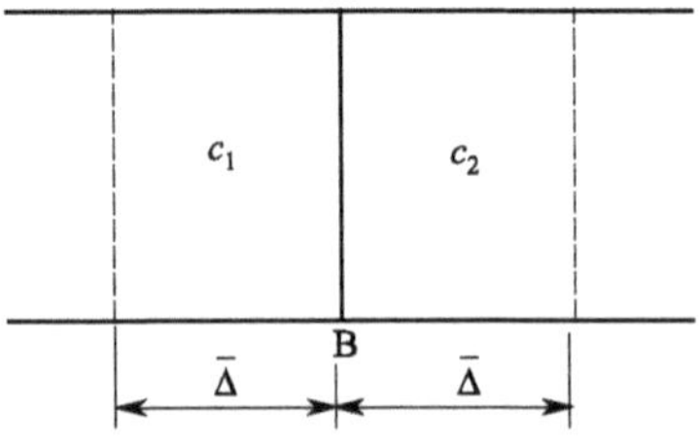

FIGURE 2.2 **Brown motion and diffusion process.**

volume of left layer can move across the cross plane AB in time t, therefore, the number of particles which migrate from the left layer to the right layer equals

$$\frac{1}{2}c_1\overline{\Delta}\cdot S$$

Although Brownian motion takes place in all directions, there are only two directional projections of the displacement in horizontal ordinate, one is from left to right, the other is from right to left, their directions are opposite but the probability are equal, that is, only a half of particles move across the cross plane AB. That's why "1/2" is added in the previous equation. Similarly, the number of particles which migrate from the right layer to the left layer equals

$$\frac{1}{2}c_2\overline{\Delta}\cdot S$$

Therefore, the net migration number of particles from left to right in horizontal ordinate is

$$m=\frac{1}{2}(c_1-c_2)\overline{\Delta}\cdot S \tag{2.5}$$

Since the thickness of liquid layer is very small

$$\frac{c_1-c_2}{\overline{\Delta}}\approx-\frac{dc}{dx}$$

Therefore, Eq. (2.5) becomes

$$m=-\frac{1}{2}\left(\overline{\Delta}\right)^2\frac{dc}{dx}S \tag{2.6}$$

Comparing Eq. (2.6) with (2.1), we have

$$\frac{1}{2}\left(\overline{\Delta}\right)^2=Dt$$

that is

$$\overline{\Delta}=(2Dt)^{\frac{1}{2}} \tag{2.7}$$

which is Einstein Brownian motion equation. It relates Brown motion with diffusion. In addition, Einstein deduced the equation for diffusion coefficient, as shown in the following:

$$D=\frac{K_\mathrm{B}T}{f} \tag{2.8}$$

where K_B is Boltzmann constant, T thermodynamics temperature, f frictional coefficient, which is the resistance force to the particle moving at unit velocity in dispersion medium. For a spherical particle

$$f = 6\pi\eta r \tag{2.9}$$

Eq. (2.9) is Stocks law, where η is viscosity of dispersion system, and r is the radius of the particle.

Example 2.2 Brownian Motion

The mean diameter of the spherical particle of a sol is 4.2 nm. Suppose the viscosity of the sol equals that of water, calculate (1) the diffusion coefficient of the particle of this sol; (2) the displacement of the particles in *x*-axis direction and in 1 s caused by Brownian motion.

Solution

1. $D = \frac{K_B T}{f} = \frac{K_B T}{6\pi\eta r} = \frac{1.38\times10^{-23}\times298}{6\times3.14\times0.89\times10^{-3}\times2.1\times10^{-9}}$
   $= 1.16\times10^{-10}\ (\mathrm{m^2\cdot s^{-1}})$

2. $\bar{\Delta} = (2Dt)^{\frac{1}{2}} = (2\times1.16\times10^{-10})^{\frac{1}{2}}$
   $= 1.52\times10^{-5}(\mathrm{m}) = 1.52\times10^{-2}(\mathrm{mm})$

2.3 APPLICATION OF DIFFUSION

2.3.1 Calculation of Radius and Molecular Weight of Spherical Particle

Substituting Eq. (2.9) into Eq. (2.8) gives

$$D = \frac{K_B T}{6\pi\eta r}$$

$$r = \frac{K_B T}{6\pi\eta D} \tag{2.10}$$

The radius obtained by Eq. (2.10) is termed hydromechanical radius. Substituting it into next equation, we will obtain the molecular weight of particle as follows.

$$M = \frac{4\pi r^3 N_A}{3\bar{V}} \tag{2.11}$$

where N_A is Avogadro constant, and $\bar{V}$ is the partial molar volume of the particles, $\mathrm{m^3\cdot g^{-1}}$.

2.3.2 Calculation of Axial Ratio of Nonspherical Particle

As mentioned previously, the asymmetry of nonspherical particle can be expressed by the axial ratio of the particle. In order to calculate the axial ratio of the particle, measure the actual molecular weight of these particles by the other method, and assume the particles were spherical. Thus from equation (2.11), we have the radius of the equivalent spherical particle as follows.

$$r = \left(\frac{3\overline{V}M}{4\pi N_A} \right)^{\frac{1}{3}}$$

Accordingly, the frictional coefficient of the equivalent spherical particle is

$$f_0 = 6\pi\eta \left(\frac{3\overline{V}M}{4\pi N_A} \right)^{\frac{1}{3}}$$

As we known, diffusion coefficient is

$$D = \frac{K_B T}{f} \tag{2.8}$$

Accordingly, the actual frictional coefficient can be obtained according to the measured diffusion coefficient D as follows.

$$f = \frac{K_B T}{D} \tag{2.12}$$

Therefore, the frictional coefficient ratio f/f_0 can be calculated from the actual frictional coefficient and the frictional coefficient of the equivalent spherical particle. It is known that the value of f/f_0 is related to the axial ratio of nonspherical particle, and Perrin derived the relationship formulas between the frictional coefficient ratio f/f_0 and the axial ratio as follows.

For prolate (a > b):

$$\frac{f}{f_0} = \frac{\left[1 - \left(\frac{b}{a} \right)^2 \right]^{\frac{1}{2}}}{\left(\frac{b}{a} \right)^{\frac{2}{3}} \ln \left\{ \frac{1 + \left[1 - \left(\frac{b}{a} \right)^2 \right]^{\frac{1}{2}}}{\frac{b}{a}} \right\}} \tag{2.13}$$

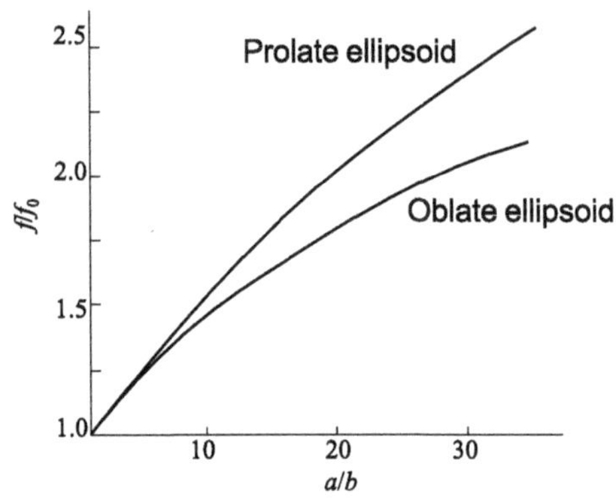

FIGURE 2.3 Result of calculation for the lower values of axial ratio by Perrin.

For oblate (b > a):

$$\frac{f}{f_0}=\frac{\left[\left(\frac{b}{a}\right)^2-1\right]^{\frac{1}{2}}}{\left(\frac{b}{a}\right)^{\frac{2}{3}}\tan^{-1}\left[\left(\frac{b}{a}\right)^2-1\right]^{\frac{1}{2}}} \tag{2.14}$$

Perrin also obtained the corresponding relationship curves for the lower values of axial ratio as shown in Fig. 2.3. Readers can consult the relationship formulas, then calculate the axial ratio values or directly obtain them from Fig. 2.3 in some condition.

2.3.3 Estimation of Amount of Solvation

In addition to asymmetry, the particles are solvated commonly, also leading to the increase of frictional coefficient. Kraemer gives

$$\left(\frac{f}{f_0}\right)_{\text{solvation}}=\left(1+\frac{W}{V\rho_0}\right)^{\frac{1}{3}} \tag{2.15}$$

where W is the amount of salvation per 1 g of particles, ρ_0 is the density of solvent, and $\overline{V}$ is the specific volume of solute. Combining these two effects gives

$$\left(\frac{f}{f_0}\right)_{\text{actual}}=\left(\frac{f}{f_0}\right)_{\text{solvation}}\left(\frac{f}{f_0}\right)_{\text{assymetry}} \tag{2.16}$$

Suppose $\left(\frac{f}{f_0}\right)_{\text{solvation}} = 1$, the maximum axial ratio can be obtained from Eq. (2.16), and suppose $\left(\frac{f}{f_0}\right)_{\text{assymetry}} = 1$, the maximum amount of solvation can be obtained from Eq. (2.16) too.

Example 2.3 Solvation and Ellipticity

The diffusion coefficient of the human hemoglobin molecule at 20°C is $6.9 \times 10^{-11}\text{m}^2 \cdot \text{s}^{-1}$. Use this value to determine f for this molecule. Suppose the particle mass $m = 1.03 \times 10^{-22}$ kg per particle, and density $d = 1.34$g per cm^3, Evaluate f_0 for hemoglobin. Indicate the possible states of salvation and ellipticity that are compatible with the f/f_0 ratio.

Solution

According to Eq. (2.12)

$$f = \frac{K_B T}{D} = \frac{1.38 \times 10^{-23} \times 293}{6.9 \times 10^{-11}} = 5.86 \times 10^{-11}\,\text{kg} \cdot \text{s}^{-1}$$

Dividing the particle mass by its density gives 7.69×10^{-26} m^3 as the particle volume. The radius of an unsolvated sphere of this volume is given by

$$r = \left(\frac{3V}{4\pi}\right)^{\frac{1}{3}} = 2.64 \times 10^{-9}\,\text{m}$$

Using the viscosity of water, 0.01P (0.001 Pa·s), the friction factor f_0 is calculated by Eq. (2.9):

$$f_0 = 6\pi\eta r = 6\pi \times 10^{-3} \times 2.64 \times 10^{-9} = 4.98 \times 10^{-11}\,\text{kg} \cdot \text{s}^{-1}$$

thus

$$\frac{f}{f_0} = \frac{5.86 \times 10^{-11}}{4.98 \times 10^{-11}} = 1.18$$

According to Eq. (2.16), if $\left(\frac{f}{f_0}\right)_{\text{assymetry}} = 1$, the particle will be a sphere, and $\left(\frac{f}{f_0}\right)_{\text{solvation}} = 1.18$. Using Eq. (2.15), we obtain that 1 g of hemoglobin has 0.48 g of salvation water. If $\left(\frac{f}{f_0}\right)_{\text{solvation}} = 1$, the particle will be unsolvated and $\left(\frac{f}{f_0}\right)_{\text{assymetry}} = 1.18$. This corresponds to an axial ratio b/a = 0.24 according to Eq. (2.13), if the particle is a prolate ellipsoid, or to $b/a = 4.0$ according to Eq. (2.14), if the particle is oblate ellipsoid.

2.4 ROLE OF DIFFUSION THEORY IN WATER TREATMENT AND NATURAL WATER

Mass transfer is of great importance in water treatment. Diffusion is one of the important forms of mass transfer. The theories and equations of diffusion are wildly used in water treatment. The most typical example is its application in perikinetic flocculation which is caused by Brownian motion of particles. The Fick's first law was successfully used to derive the kinetic equation of perikinetic flocculation, as shown in Section 7.6.1 of this book. The theory of diffusion is also the scientific basis of the double membrane model of gas transfer in water environment. In this double membrane model, Fick's first law and Henry's law were applied to establish the basic equations of gas-liquid transfer, which is of great importance for the study of the migration and conversion of gases in water, such as oxygen or carbon dioxide, etc. In addition, Streeter and Phelps derived the basic equations of BOD–DO coupling model for one-dimensional and steady state river according to Fick's second law and mass conservation law, which is widely used to determine the environmental capacity of water body. Readers can consult the books on other aspects of environmental chemistry if they are interested in these areas.

Chapter 3

Sedimentation

Chapter Outline

3.1 SEDIMENTATION IN THE GRAVITY FIELD

The separation process of particle from fluid by gravity is defined as sedimentation in the gravity field. Suppose the gravity force applied to the particles is *F*, then we have

$$F = \phi(\rho - \rho_0)g \tag{3.1}$$

where ϕ is the volume of particles, ρ and ρ_0 are the densities of particles and media, respectively, and g is the gravitational acceleration. It shows that particles will settle downward when $\rho > \rho_0$. There must be friction between particle and medium when the sedimentation takes place. If the settling velocity is not very high, this friction is proportional to the settling velocity v. Suppose the friction is F' and the friction factor is f, then we have

$$F' = f v \tag{3.2}$$

F' increases with the increase in the velocity of particles and quite rapidly (μs or ms) reaches the stationary state. Since the gravity force and the viscosity force are equal under stationary state conditions, Eqs. (3.1) and (3.2) may be equated to show

$$\phi(\rho - \rho_0)g = f v \tag{3.3}$$

where v is the velocity of a settling particle during much of its fall and is known as the stationary settling velocity. For a spherical particle, the friction factor is given by Stokes's law

Colloid and Interface Chemistry for Water Quality Control. http://dx.doi.org/10.1016/B978-0-12-809315-3.00003-7

$$f = 6\pi\eta r \tag{3.4}$$

Therefore, Eq. (3.3) becomes

$$\frac{4}{3}\pi r^3(\rho - \rho_0)g = 6\pi\eta r v$$

thus

$$v = \frac{2r^2}{9\eta}(\rho - \rho_0)g \tag{3.5}$$

This is the equation of particle sedimentation in a gravity field. It indicates that

1. The settling velocity of a particle is proportional to the particle radius squared, which means the velocity strongly depends on the size of the particle. This relationship is the basis for determining the particle size distribution of the suspension system.
2. The sedimentation process can be controlled by adjusting the difference of density between particle and medium.
3. The settling velocity of a particle is inversely proportional to the viscosity of the system; therefore, the sedimentation process can be accelerated or slowed down by changing the viscosity of the medium.

Example 3.1 Sedimentation in the Gravity Field

One small glass sphere of 1 μm in diameter settles downward in water at 20°C. If the density of the glass sphere is 2.6, how much time is needed for 1 cm of settling?

Solution

According to Eq. (3.5):

$$v = \frac{2r^2}{9\eta}(\rho - \rho_0)g = \frac{2\times(0.5\times10^{-6})^2}{9\times0.001}\times(2.6-1)\times1000\times9.8 = 871\times10^{-9}$$
$$= 8.71\times10^{-5}(\mathrm{cm\cdot s^{-1}})$$

$$t = \frac{1}{v} = \frac{1}{8.71\times10^{-5}} = 0.1148\times10^{-5}(\mathrm{s}) = 3.189(\mathrm{h})$$

3.1.1 Measurement of the Distribution of Particle Sizes by Sedimentation Analysis

Suppose a balance pan is positioned at some appropriate location below the surface of a dispersion system, as shown in Fig. 3.1. When sedimentation occurs, the settling material collects on the balance pan. The total weight P of the

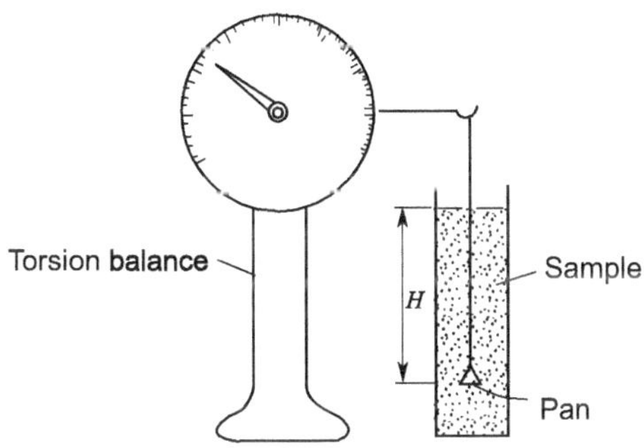

FIGURE 3.1 Settling balance.

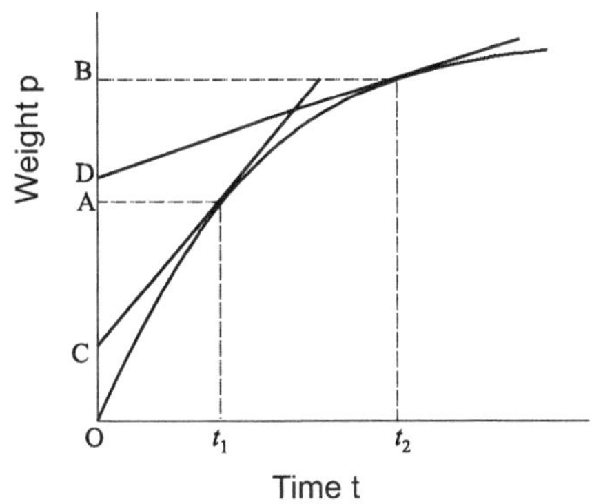

FIGURE 3.2 Settling curve.

material which has collected on the pan is measured at various times by noting the displacement of hand of torsion balance. The result is shown in Fig. 3.2.

Let the weight of materials accumulated on the pan at time t_1 be P_1, and then group the particles into two fractions according to particle size. The first is the group of particles with radii greater than a certain size, which are large enough to have this fraction of particles completely fall through the full distance between the dispersion surface and the pan. The second is the group of particles with radii smaller than a certain size, which are not large enough to have this fraction of particles completely fall through the full distance between the dispersion surface and the pan. Some of them have fallen on the pan, but some of them only fall through a part of the distance; that is, they are still on the way to landing on the pan. For example, if the distance between the pan and the surface is 30 cm, and t_1 is 300 s, all particles with the settling velocity greater than or equal to 0.1 $\text{cm} \cdot \text{s}^{-1}$ have fallen out completely, and if $\rho = 31\ \text{kg} \cdot \text{dm}^{-3}$, $\rho_0 = 1\ \text{kg} \cdot \text{dm}^{-3}$, and $\eta = 0.001\ \text{Pa} \cdot \text{s}$, we can determine from Eq. (3.5) that their radii $r \geq r_1 = 1.52 \times 10^{-3}$ cm. Considering the particles with $r \leq r_1 = 1.52 \times 10^{-3}$ cm, we know some of them have fallen on the pan, but some of them are on the way to falling. The contributions of the aforementioned two fractions to P_1 can be calculated as follows: the quantity *dP/dt* at any instant t is the slope of the cumulative curve at a particular point. It represents the rate of deposition of particles of the second fraction that have not settled out completely during the time *t*. Since the particles in this size range have been settling at the same rate *dP/dt* from the beginning of the experiment, the weight of the particles in this size range that have collected on the pan until time t is given by $t_1(dP/dt)$. The other contribution is made by the first fraction of particles that have already settled out completely by time *t*. Let the weight of the particles in this size range on the pan be S_1; the total weight *P* at any time *t* is given by

$$P_1 = S_1 + t_1 \frac{dP}{dt} \tag{3.6}$$

First, let us consider any point (t_1, OA) and draw a tangent line through this point, which meets P axis at point C, then we have $AC = (dP/dt)t_1$. It can be seen that OA = P and OC = OA−AC = S_1; therefore, intersect OC represents the weight of particles in a size range that have already settled out completely at time t_1, that is, the total weight of particles with radii greater than r_1. Similarly, intersect OD represents the weight of particles in a size range that have already settled out completely at time t_2; that is, the total weight of particles with radii greater than r_2. Therefore, the weight of particles in the range of $r_1 \sim r_2$ can be obtained by OD–OC. Dividing this difference by the total weight of all particles gives the distribution of particle size.

3.1.2 Sedimentation Equilibrium and Altitude Distribution Law

Sedimentation in the gravity field causes the concentration differences of particles from the surface to the bottom of the dispersion system. The particle concentration increases with the increase in the depth of the dispersion system. These differences will lead the diffusion of particles from the bottom to the surface. The diffusion caused is in the direction opposite to the settling, therefore, becomes the resistance to the settling. When the diffusion velocity equals the settling velocity, the dispersion system reaches the equilibrium state, which is referred to as sedimentation equilibrium. In 1910, Perrin derived an equation for sedimentation equilibrium as follows.

Suppose a dispersion system contained in a cylinder with surface area of A reaches the equilibrium state as shown in Fig. 3.3.

Let the radius of spherical particles be *r*, the density of particles and media be ρ and ρ_0 respectively, the number concentration of particles be n_1 at the height x_1 and n_2 at height x_2, and the gravitational acceleration be *g*, then the gravitational force applied to the particles in the liquid layer with thickness *dx* is given by

$$nAdx\frac{4}{3}\pi r^3(\rho - \rho_0)g \tag{3.7}$$

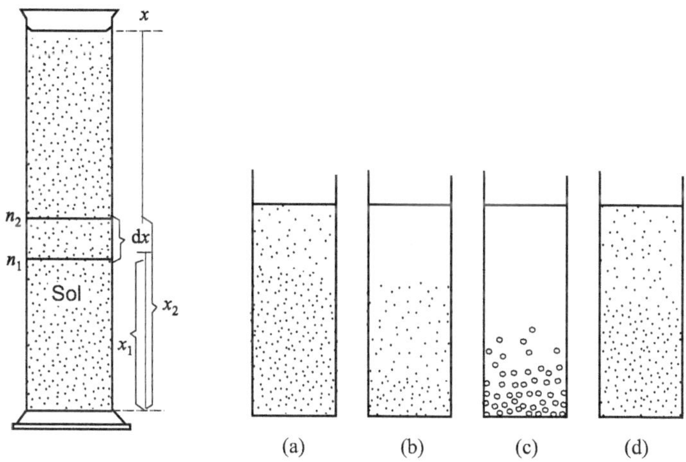

FIGURE 3.3 Sedimentation equilibrium in gravity field. (a–c) Sedimentation equilibrium of monodispersed systems with different dispersion degrees. (d) Sedimentation equilibrium of polydispersed systems consisting of (a), (b), and (c).

The diffusion force of particles is considered to be equal to the osmotic force, which makes dispersion medium flow to the region of higher concentration from the region of lower concentration and is given by

$$-\mathrm{A}\,d\pi \tag{3.8}$$

where π is osmotic pressure, $d\pi$ is the micro change of osmotic pressure with the micro change in the altitude, and the minus indicates that the diffusion force is in the opposite direction of the gravitational force. According to van't Hoff equation for osmotic pressure (see Section 4.1), we have

$$-Ad\pi = -ARTdc = -ART\frac{dn}{N_\mathrm{A}} \tag{3.9}$$

where N_A is the Avogadro constant. When the dispersion system reaches the equilibrium state at constant temperature, we have

$$-ART\frac{dn}{N_\mathrm{A}} = nAdx\frac{4}{3}\pi r^3(\rho-\rho_0)g \tag{3.10}$$

Integrating Eq. (3.10) gives

$$\ln\frac{n_2}{n_1} = -\frac{N_\mathrm{A}}{RT}\times\frac{4}{3}\pi r^3(\rho-\rho_0)(x_2-x_1)g \tag{3.11}$$

or

$$\frac{n_2}{n_1} = \exp\left[-\left(\frac{N_\mathrm{A}}{RT}\right)\frac{4}{3}\pi r^3(\rho-\rho_0)(x_2-x_1)g\right] \tag{3.12}$$

Eq. (3.12) is referred to as altitude distribution law. It shows that there is a concentration gradient in the dispersion system that is in the sedimentation equilibrium state. If the mass of particles is greater, the particle concentration will quickly decrease with the increase in altitude, and more particles will concentrate at the bottom of the system. If the mass of particles is smaller, the particle concentration will slowly decrease with the increase in altitude, and the distribution of particles is relatively uniform compared with the former; that is, the higher the altitude, the more the smaller particles, and the lower the altitude, the more the larger particles. It should be noted that the system containing larger particles reaches sedimentation equilibrium quicker, but the system containing smaller particles needs more time to reach sedimentation equilibrium. After the system containing smaller particles reaches sedimentation equilibrium, the smaller particles stop settling; as a result, the water cannot be clarified by sedimentation.

Example 3.2 Sedimentation Equilibrium and Altitude Distribution Law

A golden sol has reached sedimentation equilibrium in a gravitational field at 20°C. The number of particles in a unit volume at an altitude is 166. Calculate the number of particles in a unit volume at an altitude that is 10^{-4} m lower than the previous altitude. The densities of particles and water are 19,300 kg·m^{-3} and 1000 kg·m^{-3}, respectively.

Solution

$$\ln\frac{n_2}{n_1} = -\frac{N_A}{RT}\times\frac{4}{3}\pi r^3(\rho-\rho_0)(x_2-x_1)g$$

$$\ln\frac{n_2}{166} = -\frac{6.02\times10^{23}}{8.314\times298}\times\frac{4}{3}\times3.14\times(3\times10^{-8})^3\times(19300-1000)\times(-10^{-4})\times9.8$$

$n_2 = 272.$

3.2 SEDIMENTATION IN CENTRIFUGAL FORCE FIELD

By employing centrifugal forces instead of gravity, the application of sedimentation can be extended to the study of colloidal systems. There are two methods for studying the sedimentation in centrifugal force fields: one is velocity, the other is equilibrium.

3.2.1 Velocity Method

According to the physical principle, in centrifugal force fields the driving force on a suspended molecule or particle is

$$M(1-\bar{V}\rho_0)\omega^2 x \quad (3.13)$$

where ω is the angular velocity, x is the distance of the particle from the axis of rotation, M is the molecular weight, $\overline{V}$ is the specific volume, and ρ_0 is the density of the medium. The frictional resistance of the suspending medium for one mole of particles is

$$N_A f_s \frac{dx}{dt}$$

where N_A is the Avogadro constant, f_s is the friction factor, and t is the time. When the centrifugal force is equal to frictional resistance

$$N_A f_s \frac{dx}{dt} = M(1-\overline{V}\rho_0)\omega^2 x \tag{3.14}$$

The sedimentation coefficient is defined as

$$S = \frac{dx}{dt}\frac{1}{\omega^2 x} \tag{3.15}$$

Solving Eq. (3.14) gives

$$M = \frac{N_A f_s}{1-\overline{V}\rho_0} \cdot S \tag{3.16}$$

If the friction factor equals the frictional coefficient in the diffusion experiment, that is, $f_s = f_d$, and combining Eqs. (3.16) and (2.8) gives the molecular weight, then

$$\begin{aligned} M &= \frac{N_A f_s}{1-\overline{V}\rho_0} \cdot S = \frac{N_A K_B T f_d}{K_B T(1-\overline{V}\rho_0)} \cdot S = \frac{RT}{\frac{K_B T}{f_d}(1-\overline{V}\rho_0)} \cdot S \\ &= \frac{RT}{D(1-\overline{V}\rho_0)} \cdot S \end{aligned} \tag{3.17}$$

where K_B is the Boltzmann constant. Multiplying K_B by Avogadro constant N_A gives the gas constant R. Eq. (3.17) is referred to as the Svedberg formula, which is the basis of measuring molecular weight by the centrifugal sedimentation method. To obtain molecular weight, you should measure sedimentation coefficient first. From Eq. (3.15), we have

$$S = \frac{dx}{dt}\frac{1}{\omega^2 x} = \frac{d\ln x}{\omega^2 dt}$$

$$dt = \frac{d\ln x}{\omega^2 S}$$

$$\int_{t_1}^{t_2} dt = \frac{1}{\omega^2 S}\int_{x_1}^{x_2} d\ln x$$

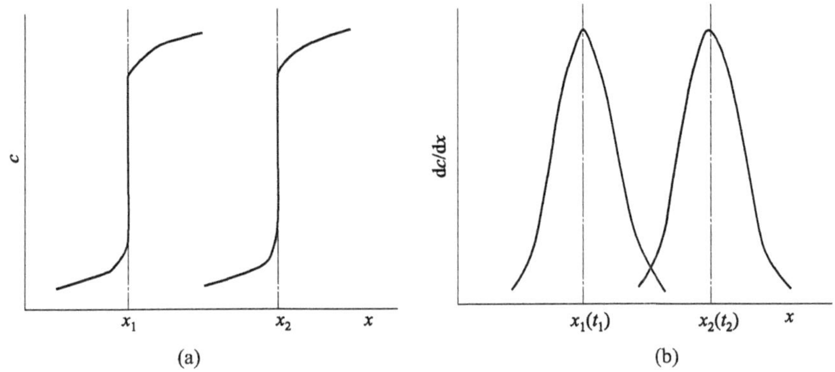

FIGURE 3.4 Measuring the sedimentation coefficient by the refractive index method. (a) The concentration sharply changes at the interface area; (b) Two peaks indicate the position of interface x_1 and x_2 at time t_1 and time t_2, respectively.

$$t_2 - t_1 = \frac{\ln \frac{x_2}{x_1}}{\omega^2 S}$$

$$S = \frac{\ln \frac{x_2}{x_1}}{\omega^2 (t_2 - t_1)} \tag{3.18}$$

In the experiment, the interface between solid and liquid forms during the process of practical centrifuge and gradually moves far away from the axis of rotation. The concentration of particles is measured by the refractive index method at different distances x, and the corresponding curve is obtained and converted to the curve of dc/dx to x as shown in Fig. 3.4.

Fig. 3.4a shows the concentration sharply changes at the interface area. In Fig. 3.4b, two peaks indicate the position of interface x_1 and x_2 at time t_1 and time t_2, respectively. Substituting them into Eq. (3.18) yields the sedimentation coefficient, thus we can obtain molecular weight from Eq. (3.17).

Example 3.3 Sedimentation in the Centrifugal Force Field: Velocity Method

What should be the speed of a centrifuge so that the interface associated with the sedimentation of a particle of molecular weight 6000 $g \cdot mol^{-1}$ moves from r_1 = 6.314 cm to r_2 = 6.367 cm in 10 min? The density of the particle and the medium are 0.728 and 0.998 $g \cdot cm^{-3}$, respectively, and the friction factor of the molecule is 5.3×10^{-11} $kg \cdot s^{-1}$

Solution

According to Eq. (3.14)

$$N_A f_s \frac{dx}{dt} = M(1 - \bar{V}\rho_0)\omega^2 x$$

If the volume of the particle is V, and the density ρ, we have

$$N_A f_s \frac{dx}{dt} = (M - M\bar{V}\rho_0)\omega^2 x$$

$$N_A f_s \frac{dx}{dt} = V(\rho - \rho_0)\omega^2 x$$

$$\frac{N_A f_s \frac{dx}{dt}}{\rho} = V\left(1 - \frac{\rho_0}{\rho}\right)\omega^2 x$$

$$\frac{f_s \frac{dx}{dt}}{\omega^2 x} = \frac{M}{N_A}\left(1 - \frac{\rho_0}{\rho}\right)$$

$$S = \frac{M}{N_A f_s}\left(1 - \frac{\rho_0}{\rho}\right)$$

Since
$M = 60{,}000\ \text{g}\cdot\text{mol}^{-1} = 60\ \text{kg}\cdot\text{mol}^{-1}$
Sedimentation coefficient is given by

$$S = \left(\frac{60}{6.02\times10^{23}\times5.3\times10^{-11}}\right)\left[1 - \left(\frac{0.998}{0.728}\right)\right] = 6.98\times10^{-13}s$$

From Eq. (3.18), we have

$$\omega^2 = \frac{\ln\left(\frac{x_2}{x_1}\right)}{S(t_2 - t_1)} = 1.997\times10^7 S^{-2}$$

$$\omega = 4.47\times10^3\ rad.S^{-1}$$

Dividing by 2π gives
$\omega = 711\ \text{r}\cdot\text{s}^{-1}$

3.2.2 Equilibrium Method

The sedimentation caused by centrifugal force generally leads the concentration gradient of particles in the centrifugal force field, just as in the gravitational field. This concentration gradient again causes the diffusion of particles at the same time. The diffusion caused is in the direction opposite of the sedimentation, from the region of higher concentration to the region of lower concentration. When the diffusion velocity equals the settling velocity,

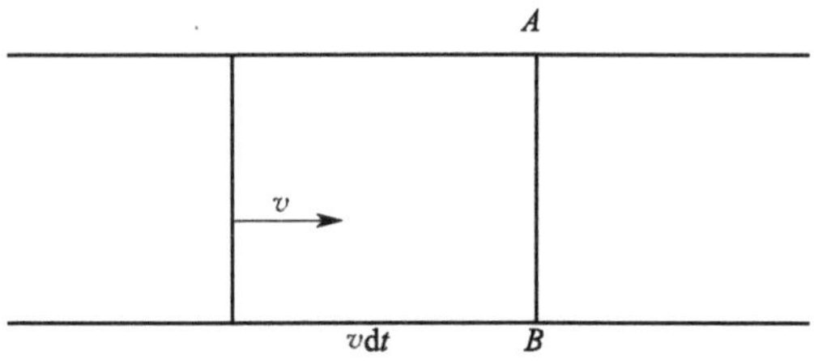

FIGURE 3.5 A particle moving across the cross section.

the system reaches the equilibrium state. According to Eq. (3.14), the settling velocity of particles is

$$v = \frac{M(1-\bar{V}\rho_0)\omega^2 x}{N_A f_s}$$

Let the cross section of a dispersion system be AB, and its area be 1 unit. Suppose a liquid layer is on the left of AB and its thickness is *vdt*, where *v* is settling velocity and t is time, as shown in Fig. 3.5.

Since the area of AB equals 1, *vdt* is equal to the volume of the liquid layer. Therefore, the number of particles moving across AB in *dt* time is

$$dm_s = vcdt = \frac{M(1-\bar{V}\rho_0)\omega^2 x}{N_A f_s} cdt \tag{3.19}$$

According to Fick's first law, the diffusion velocity is

$$dm_d = -D\frac{dc}{dx}dt = -\frac{K_B T}{f}\frac{dc}{dx}dt = -\frac{RT}{N_A f}\frac{dc}{dx}dt \tag{3.20}$$

In the equilibrium state, Eq. (3.19) is equal to Eq. (3.20)

$$\frac{M(1-\bar{V}\rho_0)\omega^2 x}{N_A f_s} cdt = -\frac{RT}{N_A f}\frac{dc}{dx}dt$$

thus

$$M = \frac{RTd\ln c}{(1-\bar{V}\rho_0)\omega^2 x dx} \tag{3.21}$$

Integrating Eq. (3.21) gives

$$M = \frac{2RT\ln\left(\frac{c_2}{c_1}\right)}{(1-\bar{V}\rho_0)\omega^2(x_2^2 - x_1^2)} \tag{3.22}$$

The molecular weight will be obtained if x_1, x_2, and c_1, c_2 are measured respectively and substituted into Eq. (3.22).

Example 3.4 Sedimentation in Centrifugal Force Field–equilibrium Method

A dispersion system in the centrifugal force field reached equilibrium state at 298 K and in 39 h. The rotation speed of centrifuge $n = 8700\ \mathrm{r \cdot min^{-1}}$, the density of medium $\rho_0 = 1.008 \times 10^3\ \mathrm{kg \cdot m^{-3}}$, and the specific volume of hemoglobin $\bar{V} = 0.740 \times 10^{-3}\ \mathrm{m^3 \cdot kg^{-1}}$. Let the concentration of hemoglobin be c, and the distances from axis of rotation be x. If $c_1 = 0.832\%$, $x_1 = 4.46 \times 10^{-2}$ m and $c_2 = 0.930\%$, $x_2 = 4.51 \times 10^{-2}$ m, respectively. Calculate the molecular weight of hemoglobin.

Solution

The angular velocity is given by

$$\omega = \frac{2\pi n}{60} = 2 \times 3.14 \times \frac{8700}{60} = 910\left(S^{-1}\right)$$

$$M = \frac{2RT\ln\left(\frac{c_2}{c_1}\right)}{\left(1-\bar{V}\rho_0\right)\omega^2\left(x_2^2 - x_1^2\right)} = \frac{2\times 8.31\times 293\times \ln\left(\frac{0.930}{0.832}\right)}{\left(1-0.749\times 10^{-3}\times 1.008\times 10^3\right)\times 910^2}$$
$$\times \frac{1}{4.51^2\times 10^{-4} - 4.46^2\times 10^{-4}} = 59.4\ (\mathrm{kg \cdot mol^{-1}})$$

3.3 APPLICATIONS OF SEDIMENTATION IN WATER TREATMENT

Sedimentation operation is one of the most important methods or units of water and wastewater treatment. The sedimentation in gravity field is commonly used as the primary sedimentation operation or clarification operation following flocculation in the treatment process of water supply. The settling times in Table 3.1 are calculated by using Eq. (3.5). This table shows that the particle can be separated from water by natural sedimentation only if its radius is greater than

TABLE 3.1 Settling Times of Spherical Particles With Different Radii[a]

Radius of the particle (μm)	Time needed for 1 cm of sedimentation
100	0.45 s
10	0.77 min
1	1.25 h
0.1	125 h

[a]*Temperature, 20°C; particle density, 2 kg dm^{-3}*

10 μm. But if the radius of particles is smaller than 10 μm, the settling velocity is too small to have the effect in practice; for example, the time needed for 1 cm of settling reaches up to 1.25 h when the radius of the particle equals 1.0 μm and, therefore, cannot meet the request for the velocity of outflow of water in settling pond or sedimentation tank in a water treatment plant. This means the smaller particles should be aggregated to become the larger particles in order to be separated from water under the effect of gravity force. This aggregation of particles can be achieved by coagulation or flocculation, in which the appropriate chemical and hydrodynamics conditions are provided, thus leading to the formation of larger flocs, which will quickly settle downward.

In the winter of northern areas, the water temperature is lower, which causes higher water viscosity, thus the small settling velocity of particles according to Eq. (3.5). In addition, the hydrolysis reaction of coagulants is also slower at lower temperatures; as a result, it is more difficult to treat water in the winter than in other seasons.

There are two kinds of settling situations in the treatment of water supply. One is the free settling in which there is no interference between particles during the settling, and only the gravitational force and the diffusion force are applied to the particles. This free settling obeys Eq. (3.5). The other is the crowded settling in which the particles interfere with each other during settling, leading to a very small settling velocity. For instance, the settling of particles in the water with high turbidity belongs to such crowded settling.

The sedimentation in gravity field is one of the important procedures of the chemical precipitation method of wastewater treatment. For example, in the treatment of the wastewater containing heavy metal ions or fluorine ions, $Ca(OH)_2$ is added to the wastewater to produce precipitates (metal hydroxide or calcium fluoride), which is then followed by settling. The efficiency of chemical precipitation treatment depends on the settling effect to a large extent. The sedimentation is also used in the pretreatment or as the first settling before biochemical treatment. In some cases the sedimentation is used as the second settling after biochemical treatment to remove the microorganism and the sludge.

In sludge treatment, to reduce the volume of the sludge in order to decrease the scale of the equipment and the cost of the subsequent sludge treatment.

Sedimentation in centrifugal force fields is commonly used in the hydraulic cyclone, the hydraulic cyclone settling pond, and the centrifugal machine for sludge dewatering. Pressure-type hydraulic cyclone consists of an upper cylinder and a lower cone. The liquid to be separated is pumped into the device along a tangential direction. The tangential motion then becomes the rotational motion because of the boundary restraint, producing great centrifugal force. Then the larger particles move to the side wall, slide downward along the side wall, and are discharged at the bottom. At the same time, the clear liquid is discharged through the upper tube. As far as the hydraulic cyclone settling pond is concerned, the rotational flow is produced by varying water levels. In this equipment, the volume is larger, but the energy consumption is smaller, and the surface load is lower. Readers can consult the corresponding books or works in this area.

In recent decades, a large amount of biological and chemical sludge has been produced in water treatment plants, which makes sludge dewatering more and more important. Considering the economic cost and space occupied, people are commonly interested in the centrifuge method. There are many advantages to this method; for example, the equipment can be operated continuously, the working environment is cleaner, the area and space occupied are smaller, the flocculant consumption is less, and the request for flocs strength and flocs size are lower.

Chapter 4

Osmotic Pressure

Chapter Outline

Osmotic pressure is very common in nature; for example, by osmotic pressure, water is driven to rise from soil to the body of plants and transfer within its leaves and branches. Here we should recognize that osmotic pressure is not suitable for studying lyophobic colloidal systems. There are two reasons for this: One is that its particles aggregate and settle easily, and the other is its value measured is too low to be measured with sufficient accuracy. Although osmotic pressure is not suitable for studying lyophobic colloid systems, it is very important for studying macromolecule solutions because the macromolecule solution is thermodynamically stable and provides a measurable effect. Osmotic pressure is a colligative property, and its effect is much more significant than other colligative properties, including lowering of vapor pressure, depression of freezing point, elevation of boiling point, etc.; therefore, the research results obtained by this method are very accurate.

4.1 OSMOTIC PRESSURE OF IDEAL SOLUTIONS

Before discussing the osmotic pressure of macromolecule solutions, we should introduce the osmotic pressure of ideal solutions as the basis. Assuming there is a solution in which the interaction between solute molecules is the same as the interaction between solvent molecules and is also the same as the interaction between solute molecules and solvent molecules, such a solution is defined as an ideal solution.

Fig. 4.1 shows two liquid phases separated by a partition known as the semipermeable membrane that is permeable to the solvent and impermeable to the solute.

Colloid and Interface Chemistry for Water Quality Control. http://dx.doi.org/10.1016/B978-0-12-809315-3.00004-9

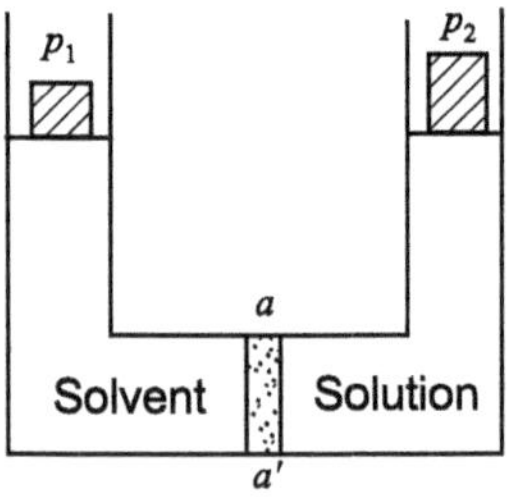

FIGURE 4.1 Schematic representation of an osmotic pressure experiment.

Let us add an ideal solution on the right side of the semipermeable membrane and its pure solvent on the left side of the semipermeable membrane in the apparatus such as that shown in Fig. 4.1. In such a case, the molecules of the solvent will permeate across the membrane from the left to the right, leading to the rise of the solution surface. This phenomenon is referred to as osmosis. The driving force that causes the osmosis is what is known as osmotic pressure of the solution, commonly represented by π.

Suppose pressure P_1 and P_2 are applied to the solvent and the solution respectively, as shown in Fig. 4.1, $P_2 > P_1$, and $P_2 - P_1 = \pi$, thus the system will reach osmotic equilibrium. In this case, the solution on the right side neither rises up nor falls down; the chemical potential of solvent on the left side is equal to the chemical potential of solvent on the right side. If P represents P_1, $P+\pi$ represents P_2, and subscript 1 represents the solvent, subscript 2 represents the solution, x represents the molar fraction, n represents the mole number, and $\overline{V}$ the specific volume, we have

$$\mu_1^0(P) = \mu_1(P+\pi, x_1) \tag{4.1}$$

$$\mu_1^0(P) = \mu_1^0(P+\pi) + RT\ln x_1 = \mu_1^0(P) + \int_P^{P+\pi} \overline{V}dp + RT\ln x_1$$

$$\mu_1^0(P) = \mu_1^0(P) + \pi\overline{V} + RT\ln x_1$$

$$-RT\ln x_1 = \pi\overline{V} \tag{4.2}$$

Since

$$\ln x_1 = \ln(1-x_2) \approx -x_2 \approx -\frac{n_2}{n_1}$$

Eq. (4.2) becomes

$$RT\frac{n_2}{n_1} = \pi\overline{V}$$

thus

$$\pi = \frac{RTn_2}{n_1 \overline{V}} = \frac{n_2}{V} RT \tag{4.3}$$

or

$$\pi = RT \frac{c}{M} \tag{4.4}$$

where V is the volume of solution, M is the molecular weight (kg mol^{-1}) of solute, and c is mass concentration (kg m^{-3}) of solute. It can be seen that Eq. (4.3) is very similar in form to the state equation of ideal gas.

4.2 OSMOTIC PRESSURE OF MACROMOLECULE SOLUTIONS

There are some deviations between macromolecule solutions and ideal solutions. These deviations can be corrected by multiple correction factors

$$\pi = RT(A_1 c + A_2 c^2 + A_3 c^3 + \cdots\cdots) \tag{4.5}$$

where A_1, A_2, A_3,...... are referred to as virial coefficients. The first virial coefficient is

$$A_1 = \frac{1}{M} \tag{4.6}$$

For dilute macromolecule solutions, Eq. (4.5) can be simplified as

$$\pi = RT(A_1 c + A_2 c^2) = cRT(A_1 + A_2 c) \tag{4.7}$$

or

$$\frac{\pi}{c} = RT\left(\frac{1}{M} + A_2 c\right) \tag{4.8}$$

This form suggests that the plot of π / c versus c should be a straight line, and from the intercept and slop of which, we can obtain the molecular weight and the second virial coefficient. The second virial coefficient A_2 indicates the relationship between macromolecule and solvent as follows.

$A_2 > 0$: The attractive force between molecular chain segments is weaker, but the attractive force between molecular chain segment and solvent is stronger, therefore, the solvent is the fine one;

$A_2 = 0$: The attractive force between molecular chain segments is equal to the attractive force between molecular chain segment and solvent, therefore, the solution is the ideal solution;

$A2 < 0$: The attractive force between molecular chain segments is stronger, but the attractive force between molecular chain segment and solvent is weaker, therefore, the solvent is not the fine one;

The second virial coefficient A_2 changes with temperature; the temperature at which A_2 changes to zero is referred to as θ-temperature. The second virial coefficient A_2 also changes with solvent; the solvent at which A_2 changes to zero is referred to as θ-solvent. The solution which is at θ-temperature or in θ-solvent is the ideal solution.

Example 4.1 Osmotic Pressure of Macromolecule Solutions

Polyisobutylene was dissolved in benzene, forming a series of solutions. The osmotic pressures of these solutions were measured at 25°C as follows:

c (g dm^{-3})	5.0	10.0	15.0	20.0
π (Pa)	49.54	101.04	155.0	210.9

Calculate the molecular weight of this polyisobutylene.

Solution

For dilute macromolecule solutions

$$\frac{\pi}{c} = RT\left(\frac{1}{M} + A_2 c\right) \quad (4.8)$$

First, transform the data as follows:

c (kg m^{-3})	5.0	10.0	15.0	20.0
π/c (Pa m^3 kg^{-1})	9.91	10.10	10.33	10.55

Then plot this data in the form of Eq. (4.8). A plot of π/c against concentrations is shown in the following figure.

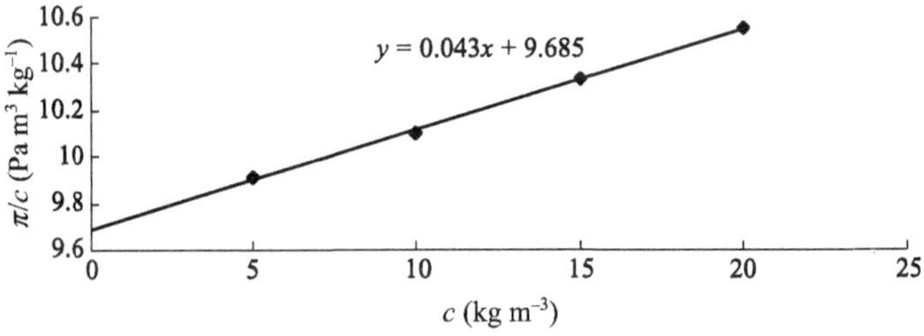

From this figure, we obtain

The intercept = π/c = 9.685 Pa m^{-3} kg^{-1}

$$M = \frac{RT}{\left(\frac{\pi}{c}\right)} = \frac{8.3145 \times 298.15}{9.685} = 255.9\ (\text{kg mol}^{-1})$$

4.3 DONNAN EQUILIBRIUM AND OSMOTIC PRESSURE

In this section, let us discuss the osmotic pressure of polyelectrolytes. The macromolecule that is able to ionize to form macro ions and small ions is defined as a polyelectrolyte. Commonly, one molecule of polyelectrolyte produces one macro ion and several small ions. The membrane is assumed to be permeable not only to the solvent but also to the small ions ionized from the polyelectrolyte, but not to the macro ion. At equilibrium the low molecular weight ions will be found on both sides of the membrane, but not in equal concentration because of the presence of the macro ions on one side of the membrane. This equilibrium is called the Donnan equilibrium. For instance, let the protein molecule be Na_zP, which ionizes as follows:

$$Na_Z\,P \rightarrow zNa^+ + P^{z-} \tag{4.9}$$

Fig. 4.2 shows that before diffusion takes place, the protein with concentration m_1 ionizes to form sodium ions with concentration zm_1 and macroions $P^{z-}$with concentration m_1 on the left side of the semipermeable membrane; in addition, the sodium chloride with the concentration m_2 ionizes to form sodium ions with concentration m_2 and chloride ions with concentration m_2 on the right side of the semipermeable membrane, as shown on the left of Fig. 4.2. If diffusion takes place, osmosis will attain equilibrium, x mole of Na^+ will move to the left side from the right side, and also x mole of Cl^- will move to the left side from the right side to maintain electric neutrality, as shown on the right of Fig. 4.2.

Since the system has reached equilibrium, the chemical potential of sodium chloride on both sides is equal

$$\mu_{NaCl}\,(\text{left}) = \mu'_{NaCl}\,(\text{right})$$

$$RT \ln a_{NaCl} = RT \ln a'_{NaCl}$$

$$a_{Na^+} \cdot a_{Cl^-} = a'_{Na^+} \cdot a'_{Cl^-}$$

where a is the activity of ions, substituting concentration for activity gives

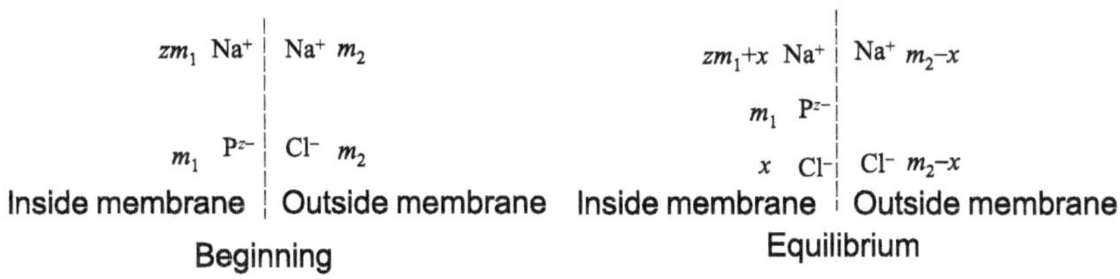

FIGURE 4.2 Donnan equilibrium.

$$(zm_1 + x)x = (m_2 - x)^2$$

where m is the mole concentration of ions, thus

$$x = \frac{m_2^2}{zm_1 + 2m_2} \tag{4.10}$$

and

$$\frac{[\mathrm{NaCl}]_{\mathrm{right}}}{[\mathrm{NaCl}]_{\mathrm{left}}} = \frac{m_2 - x}{x} = \frac{m_2}{x} - 1 = m_2 \cdot \frac{zm_1 + 2m_2}{m_2^2} - 1 = 1 + \frac{zm_1}{m_2} \tag{4.11}$$

Therefore, we have the following:

1. The concentration of sodium chloride on the two sides of the semipermeable membrane are different. This concentration difference causes osmotic pressure. The higher the charge of polyelectrolytes, the higher the osmotic pressure.
2. Nearly all of the sodium chloride is on the right side of the semipermeable membrane when $m_1 \gg m_2$.
3. Sodium chloride is equally distributed on both side of the semipermeable membrane when $m_2 \gg m_1$.

When measuring the molecular weight of polyelectrolytes, the effect of Donnan equilibrium should be eliminated by these methods:

1. Increasing the concentration of diffusible electrolytes;
2. Decreasing the concentration of polyelectrolytes;
3. Adjusting the pH value of the solution to make polyelectrolytes at their isoelectric point.

Example 4.2 Donnan Equilibrium

There is a 0.1 mol dm^{-3} solution of polyelectrolyte R^+Cl^- on the left side of the semipermeable membrane, and a 0.5 mol dm^{-3} solution of sodium chloride on the right side of the semipermeable membrane. R^+ represents the macroion to which the semipermeable membrane is not permeable. Calculate the osmotic pressure of this system that has attained Donnan equilibrium at 25°C.

Solution

For 1-1 electrolyte

$$x = \frac{m_2^2}{m_1 + 2m_2} \tag{4.10}$$

The osmotic pressure is given by

$$\pi = (2m_1 - 2m_2 + 4x)RT \tag{4.12}$$

where $(2m_1-2m_2+4x)$ is the difference of concentration between two sides of the membrane as shown in Fig. 4.2. Substituting Eq. (4.10) into Eq. (4.12) gives

$$\pi=\frac{2m_1^2+2m_1m_2}{m_1+2m_2}RT=\frac{2(0.1^2+0.1\times0.5)\times10^6}{(0.1+2\times0.5)\times10^3}\times8.314\times298.15$$
$$=270.4(\mathrm{kPa})$$

4.4 MEASUREMENT OF OSMOTIC PRESSURE

4.4.1 Osmometer

The value of osmotic pressure is measured by osmometer. First, add the solution and its solvent into the cells on the right side and the left side of membrane, respectively, and then measure the difference of the height of two liquid columns after the system attains osmotic equilibrium; thus, we will obtain the osmotic pressure. The practical osmometer should meet these requirements:

1. The establishment of equilibrium is rapid, therefore, the area of the membrane should be as large as possible, and the cross area of the liquid column should be small.
2. The effect of temperature is small, therefore, the heat capacity should be great, and the volume of the osmosis cell should be small.
3. The structure is simple and easy for the operator to replace the liquid.

Here we introduce the Fuoss–Mead osmometer. In this osmometer, the membrane is firmly clamped (and also acts as a gasket) between two carefully machined stainless steel blocks provided with channels into which small volumes of solvent and solution are introduced, as shown in Fig. 4.3. Owing to the high ratio of membrane surface area to solution volume, the approach to equilibrium is rapid. With fast membranes, using the half-sum method, measurements in well under 1 h are possible.

4.4.2 Semipermeable Membrane

In many ways the semipermeable membrane is the central feature of osmometry. It should meet the following requirements:

1. It is able to retain macromolecules, otherwise the "leak" between the two compartments would result in errors or totally invalidate the results.
2. It does not react with the solvent or macromolecule.
3. It is very permeable to the solvent, thus leading to the rapid establishment of equilibrium.

For the well-graded macromolecule of high molecular weight, it is easy to meet the aforementioned requirements. But for the macromolecule with the wide distribution of molecular weight, especially when the solution contains some

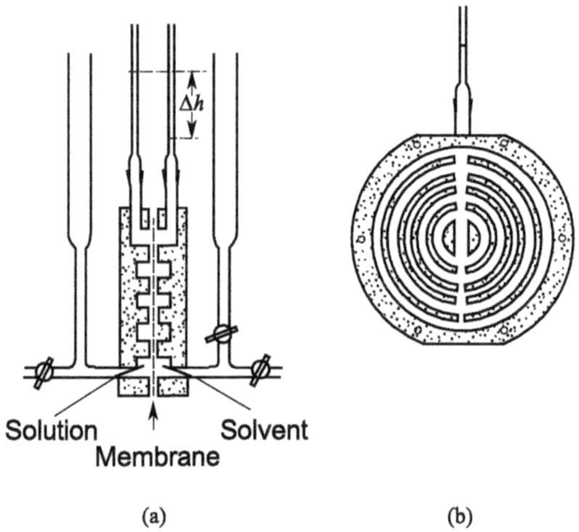

FIGURE 4.3 Schematic representation of the Fuoss–Mead osmometer. (a) Vertical cross section; (b) inner surface of each half-cell.

molecule with lower molecular weight, the "leak" will be the serious problem. In such situations, the molecules with lower molecular weight transmit across the membrane, thus leading to large errors because osmotic pressure is a colligative property; the same number of particles makes the same contribution to the osmotic pressure whether they are large or small ones. For example, there were two ungraded solutions of nitrocellulose in butyl acetate A and B; the molecular weight of the former was smaller, and the molecular weight of the latter was greater. Their osmotic pressures were measured by six kinds of semipermeable membranes of different permeable properties, respectively, and the curves of π/c against c were obtained as shown in Fig. 4.4. It can be seen that there are great differences in the results of measurement for the smaller molecular weight sample A, but there are little differences in the results of measurement for the larger molecular weight sample B.

Fig. 4.5 shows the results obtained by measuring the osmotic pressure of sample B by the same membrane as that in the previous experiment and after four times of precipitation by which 30% of macromolecules of lower molecular weight were removed. It can be seen that the results do not depend on the kinds of semipermeable membranes under these circumstances.

Generally speaking, when molecular weight is smaller than 1×10^4, various kinds of membranes are not ideal; when molecular weight is greater than 1×10^4, the measurement of osmotic pressure can provide a reliable result. But when molecular weight is greater than 1×10^6, the osmotic pressure is too small to obtain an accurate result; therefore, the appropriate range of molecular weight is $1 \times 10^4 \sim 1 \times 10^6$ for determining molecular weight by measuring osmotic pressure.

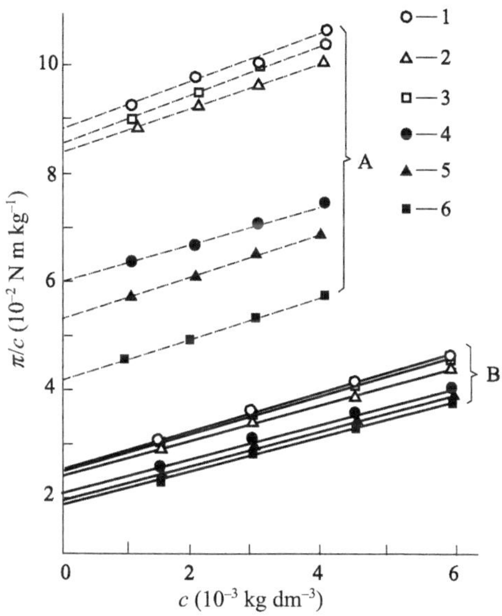

FIGURE 4.4 Osmotic pressures of solution A and B measured by different semipermeable membranes.

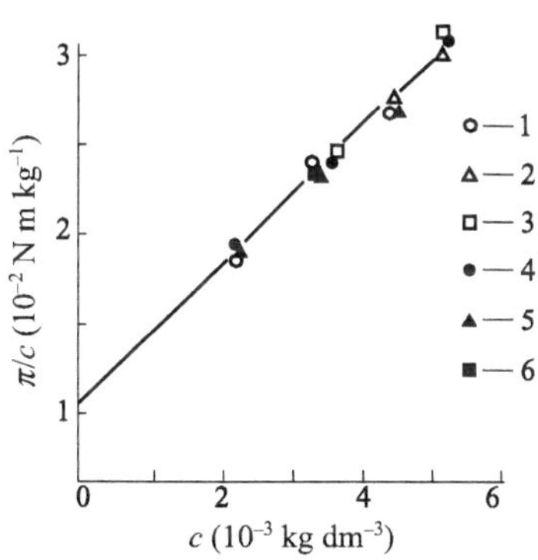

FIGURE 4.5 Osmotic pressures of solution B measured by different semipermeable membranes after gradation.

4.4.3 Method of Measurement

We have noted that osmotic equilibrium is reached slowly. In some cases as much as one week may be required for equilibrium to be achieved. To shorten this time, the half-sum method has been developed for measuring osmotic pressure in the laboratory.

Let us set the liquid level in turn at slightly (Δh) above and slightly (Δh) below the anticipated equilibrium level and then plot its position as a function of time. If the estimation of the final equilibrium level has been reasonably

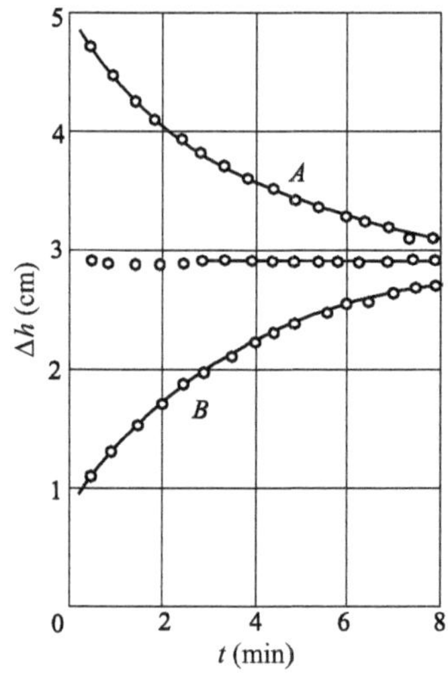

FIGURE 4.6 Approach to osmotic pressure by the half-sum method.

accurate, then the two curves will be almost symmetrical, and by plotting the half-sum as a function of time the equilibrium level can be reliably estimated after relatively short times, as shown in Fig. 4.6.

4.5 APPLICATION OF REVERSE OSMOSIS IN WATER TREATMENT

As discussed previously, the solution and solvent separated by a semipermeable membrane will be at equilibrium state only when the solution is at osmotic pressure. If the solution is under less pressure than the equilibrium osmotic pressure, solvent will flow from the pure solvent phase into the solution. If, on the other hand, the solution is under a pressure greater than the equilibrium osmotic pressure, the pure solvent will flow in the reverse direction, from the solution to the solvent phase. In the latter case, the process is referred to by the term *reverse osmosis*.

Reverse osmosis has been applied to a wide variety of systems. An application that has attracted a great deal of interest in recent years is the production of potable water from saline water. Since no phase transitions are involved as, for example, in distillation, the method offers economic feasibility in coastal regions.

Cellulose acetate seems to be the most thoroughly investigated of many possible membrane materials. Cellulose acetate membranes are capable of yielding 96 ~ 98% retention of NaCl, for example, and of delivering about 0.2 $cm^3\ s^{-1}\ atm^{-1}\ m^{-2}$ of water. This amounts to about 50 gal $day^{-1}\ ft^{-2}$at 100 atm. In recent decades, many kinds of membranes, including polyamide (PA) membrane, polypropylene (PP) membrane, and polyvinylidene fluoride (PVDF) membrane, have been developed. These new membranes have better performance in water treatment compared to traditional membranes.

When reverse osmosis is carried out to desalt saline water, the concentration of salts in the raw water will constantly increase with the production of pure water, thus leading the increase in osmosis pressure; therefore, the applied pressure is greater than the osmosis pressure of the raw water. In addition, to reach a certain production capacity, the applied pressure is much greater than the osmosis pressure of the raw water.

Currently, reverse osmosis technology has been widely applied to the production of pure water, desalination of seawater and other saline water, the treatment of some industrial wastewaters, the tertiary treatment of industrial wastewater and domestic sewages, and more.

In these applications of osmosis and reverse osmosis, it is not the solute but the membrane that makes up the colloidal system. In this case, the solid portion of the membrane would be the continuous phase, and the pores, necessarily small if the membrane is to be effective, would be the "dispersed" phase. According to this point of view, colloid and interface chemists become interested in numerous aspects of membrane technology.

Chapter 5

Optical Properties

Chapter Outline

5.1 LIGHT SCATTERING OF COLLOID SYSTEMS

When a beam of light is directed at a colloid system, the intensity of light can be measured or observed from any other direction, including the front and back, as shown in Fig. 5.1. This phenomenon is termed light scattering and is very common in the natural world, for example, the essence of the Tyndall effect is light scattering. Note that the wavelength of the light does not change before and after scattering.

The explanation for light scattering is given as follows. Light is one kind of electromagnetic wave. When an electromagnetic wave falls on a small ($<0.05 \sim 0.1\lambda$) particle, oscillating dipoles in the particle will be induced. The particle then acts as a secondary source, emitting the electromagnetic wave of the same wavelength as the incident light. If the medium is optically heterogeneous, the electromagnetic waves emitted cannot offset one another because of their different amplitudes, thus leading to the scattered light. Therefore, one of the necessary conditions for producing light scattering is the optical heterogeneity of the medium. This optical heterogeneity of medium is commonly caused by introducing colloidal particles into the medium or the density fluctuations of medium, which are resulted in by thermal motion. The other necessary condition for producing light scattering is that the diameter of the particle is less than the wavelength of incident light. If the diameter of the particle is greater than the wavelength of incident light, light reflex will take place. In such a case, the angle of incidence equals the angle of reflection.

Colloid and Interface Chemistry for Water Quality Control. http://dx.doi.org/10.1016/B978-0-12-809315-3.00005-0

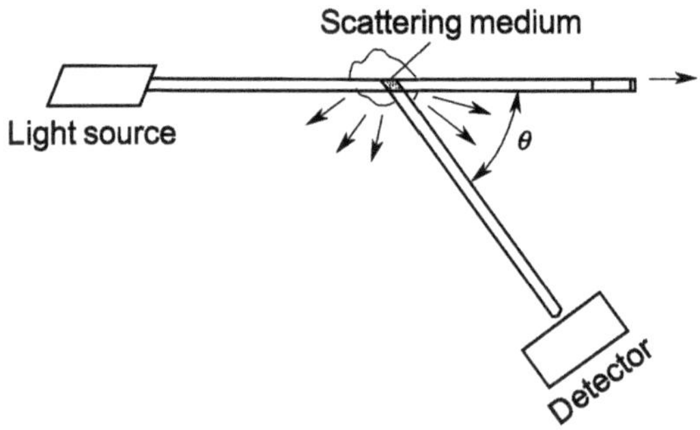

FIGURE 5.1 **Schematic representation of light scattering.**

The Tyndall effect can be observed by ultramicroscope, in which an intense illuminating beam is used to illuminate the colloid system from its side. The particles are observed as bright specks against a dark background, and their Brownian motion is shown clearly. Ultramicroscope has extended people's vision greatly, but we are able to see only the bright speck formed by light scattering, not the particle itself. The bright speck that we see by ultramicroscope is many times larger than the true particle, and we cannot see the shape and size of the particle itself. Therefore, ultramicroscope is only able to prove the existence of the particle.

Example 5.1 Light Scattering of Colloid Systems

There are four options when answering the question: What can be used to directly obtain the size and shape of individual colloidal particle?

(1) Common microscope, (2) the Tyndall effect, (3) electron microscope and (4) ultramicroscope

Solution

The magnification of the common microscope is not great enough. By the Tyndall effect or ultramicroscope, we can only see the bright speck formed by light scattering; that is, we cannot see the particle itself. Only the electron microscope is able to directly obtain the size and shape of individual colloidal particles.

The intensity of the light scattered by colloidal systems can be measured directly by the light scattering photometer shown in Fig. 5.2. In this instrument, a

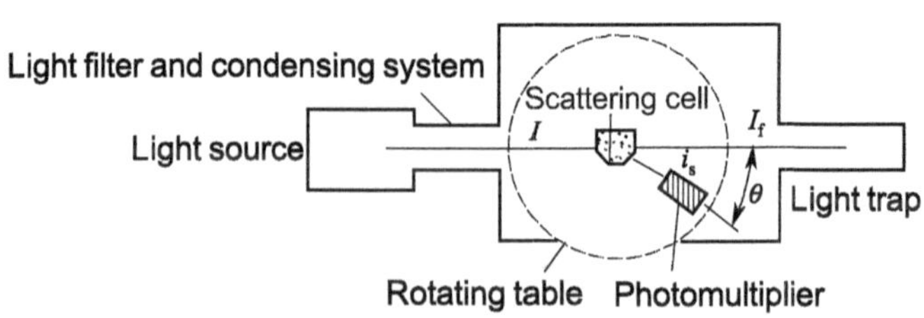

FIGURE 5.2 **Light-scattering photometer.**

beam of light emitted by a mercury lamp is changed to a collimated monochromatic light of wavelength 436 or 546 nm after it transmits across a light filter and a condensing system. This collimated monochromatic light illuminates the colloid system to produce scattered light. A detecting photocell is generally mounted on a rotating arm to permit measurement of the light scattered at different angles.

5.2 RAYLEIGH EQUATION OF LIGHT SCATTERING

Rayleigh (1871) laid the foundation of light scattering theory by applying the electromagnetic theory of light to the scattering by small, nonabsorbing (insulating), spherical particles. In his theory, the light scattering ability of the dispersion system is represented by the Rayleigh ratio

$$R_\theta = \left(\frac{i_\theta r^2}{I}\right) \tag{5.1}$$

where R_θ is defined as the Rayleigh ratio, in which r is the distance of the receptor from the scattering source; θ is the angle of observation direction to the incident beam; i_θ is the intensity of scattered light, as measured at r and θ, produced by unit volume of scattering source; and I is the intensity of incident light. Since the value of the Rayleigh ratio depends on the angle, θ is added as the subscript of R. The dimension of R_θ is m^{-1}. Eq. (5.1) shows that the greater the intensity of scattered light at constant r and I, the greater the Rayleigh ratio, that is, the greater the light scattering ability of the system. The basic assumptions for the Rayleigh equation of light scattering are as follows.

1. The size of the particle is much smaller than the wavelength of incident light, and the particle is thought as the point scattering source.
2. The concentration of colloid is very low, there are no interactions between particles, and the intensity of scattered light of the system is equal to the sum of the intensity of scattered lights of all particles.
3. The particle is isotropic, nonabsorbing, and insulating.

According to the aforementioned basic assumptions, the Rayleigh equation of light scattering is given by

$$R_\theta = \frac{i_\theta r^2}{I} = \frac{9\pi^2}{2\lambda^4}\left(\frac{n_1^2 - n_0^2}{n_1^2 + 2n_0^2}\right)^2 N_0 V^2 (1+\cos^2\theta) \tag{5.2}$$

where n_1 and n_0 are the refractive index of the dispersed phase and medium respectively; V the volume of one particle; N_0 the number of particles per unit volume of dispersion, that is, number concentration; and λ the wavelength of incident light. The Rayleigh equation shows:

1. Since scattering intensity is inversely proportional to λ^4, blue light ($\lambda \sim 450$ nm) is scattered much more than red light ($\lambda \sim 650$ nm). With

incident natural light, a scattering material will, therefore, tend to appear to be blue when viewed at right angle to the incident beam, and red when viewed from end on. This phenomenon is evident in the blue color of the sky, tobacco smoke, dilute milk, etc., and the yellowish-red of the rising and setting sun.

2. The greater the difference of the refractive index between dispersed phase and medium, the stronger the scattered light; therefore, the scattered light of sol is stronger than the macromolecule solution.
3. Since scattering intensity is proportional to V^2, the scattered light of large particles is much stronger than small particles.
4. Since scattering intensity is proportional to N_0, the higher the concentration of particle, the stronger the scattered light.

Let c be the mass concentration of particles, ρ the density of particles, then the volume of particles per unit volume of dispersion system is given by

$$N_0 V = \frac{c}{\rho} \tag{5.3}$$

Substituting Eq. (5.3) into Eq. (5.2) gives

$$R_\theta = \frac{i_\theta r^2}{I} = \frac{9\pi^2}{2\lambda^4}\left(\frac{n_1^2 - n_0^2}{n_1^2 + 2n_0^2}\right)^2 \frac{c}{\rho} V(1+\cos^2\theta) \tag{5.4}$$

Eq. (5.4) shows the Rayleigh ratio is proportional to the mass concentration of particles, which is the basis of turbidity analysis.

Example 5.2 Rayleigh Equation of Light Scattering

Give the reason why the sky appears to be blue on fine days.

Solution

The radii of the particles dispersed in the atmosphere are in the range of $10 \sim 1000$ nm; therefore, the atmosphere is a colloid dispersion system (aerosol). When the atmosphere is illuminated by sunlight, in which the wavelength of blue light is shorter than other light, including red, brown, yellow, green, etc., light scattering takes place. The scattered light produced by blue light is stronger than the other lights because scattering intensity is inversely proportional to λ^4 according to the Rayleigh equation; therefore, the sky appears to be blue.

5.3 POLARIZED COMPONENTS AND SPACE DISTRIBUTION OF SCATTERED LIGHT

From the principle of physics, we know that natural light can be resolved into two components: one is the horizontally polarized component I_H, the other is the vertically polarized component I_V. According to the theory of optics, we have

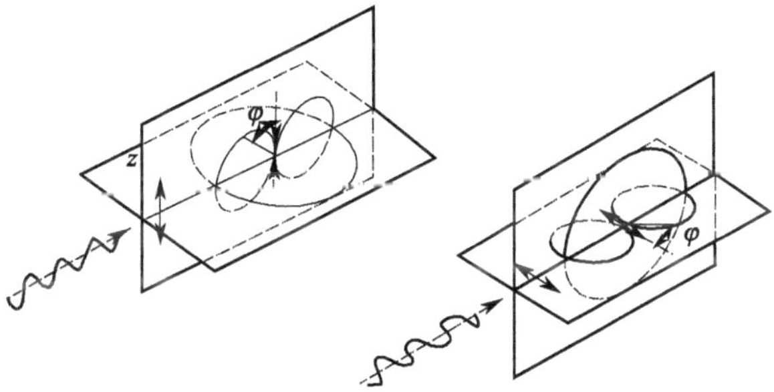

(a) (b)

FIGURE 5.3 Intensity of scattered light produced by a polarized component. (a) Vertically polarized component and (b) horizontally polarized component.

$$i_\theta \propto \sin^2 \varphi \tag{5.5}$$

where φ is the angle of the observation direction to the axis of the dipole. The second wave, produced by vertically polarized component I_V, is shown in Fig. 5.3a. Suppose an observer is on the horizontal surface, φ must equal 90 degree at any observation direction. According to Eq. (5.5), the intensity of scattered light is equal at any direction; therefore, the intensity of scattered light on the horizontal surface can be represented by a circle shape. Again, suppose an observer is on the vertical surface, φ must change from 0 to 360 degree, thus the intensity of scattered light on the vertical surface changes with square sine function. Therefore, the intensity of scattered light can be represented by a dumbbell shape. The second wave, produced by horizontally polarized component I_H, is shown in Fig. 5.3b. Similarly, we know that the intensity of scattered light on the vertical surface can be represented by a circle shape, while the intensity of scattered light on the horizontal surface can be represented by a dumbbell shape. To obtain the space distribution of scattered light, let us add the intensity of scattered light on the horizontal surface in Fig. 5.3a to the intensity of scattered light on the horizontal surface in Fig. 5.3b, or let us add the intensity of scattered light on the vertical surface in Fig. 5.3a to the intensity of scattered light on the vertical surface in Fig. 5.3b, so that we obtain the intensity of scattered light produced by unpolarized light on any surface. If we rotate this surface around the incident beam, we will obtain the space distribution of scattered light as shown in Fig. 5.4.

In conclusion, we have the following:

1. The strongest scattered light is in the direction of $\theta = 0$ degree and $\theta = 180$ degree.
2. The weakest scattered light is in the direction of $\theta = 90$ degree.
3. Since $i(\theta) = i(180\text{ degree} - \theta)$, forward scatter is equal to back scatter.
4. The scattered light is completely polarized in the direction of $\theta = 90$ degree, and unpolarized in the direction of $\theta = 90$ degree or 180 degree.

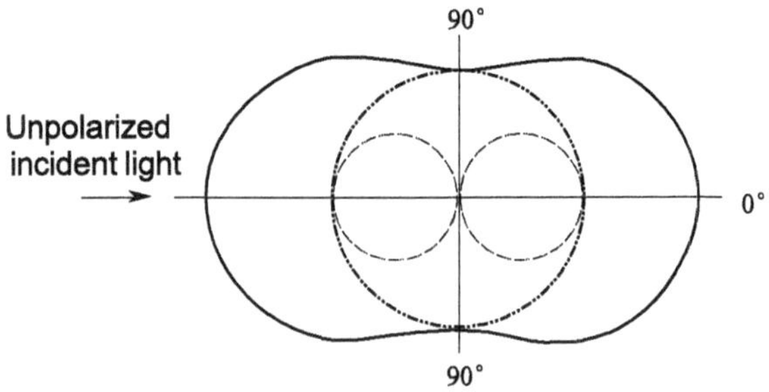

FIGURE 5.4 **The space distribution of scattered light.**

5.4 LIGHT SCATTERING OF LARGE PARTICLES

When the diameter of particles is greater than $0.05\lambda \sim 0.1\lambda$, the particles cannot be considered as the point source of scattered light. The constructive and destructive interference between scattered light waves originating from different locations on the same particle must be taken into account. Fig. 5.5 shows how this comes about. In such a case, the Rayleigh equation of light scattering does not apply to the system. Mie (1908) detailed a general quantitative theory to solve the problem. According to the Mie theory, when two incident light beams strike two different portions of the scatter, designated A and B, the light scattered from these two sites will be out of phase and display interference when observed at a larger distance compared to Δx. As shown in the figure, this interference effect may be constructive or destructive, depending on the value of θ. The destructive interference is zero for light scattered in a forward direction ($\theta = 0$ degree). Extinction will take place between waves scattered backward from the front and rear of a spherical particle of diameter $\lambda/4$ (ie, the total path difference is $\lambda/2$). The radiation envelope for such a particle will, therefore, be unsymmetrical, with more light being scattered forward than backward as shown in Fig. 5.6.

When particles of significantly different refractive index from that of the suspending medium contain a dimension greater than $\sim\lambda/4$, extinction at intermediate

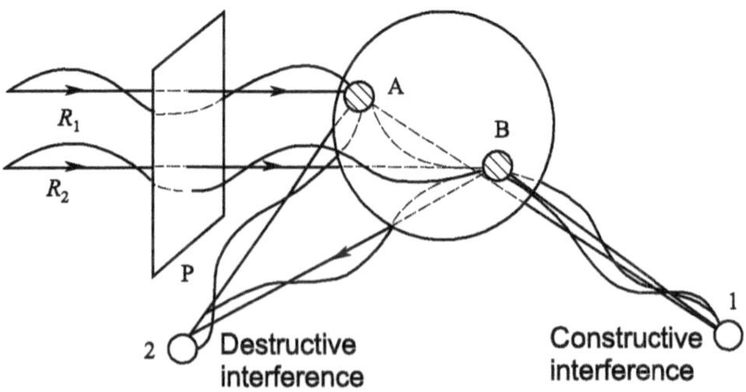

FIGURE 5.5 **Interference between scattered light.**

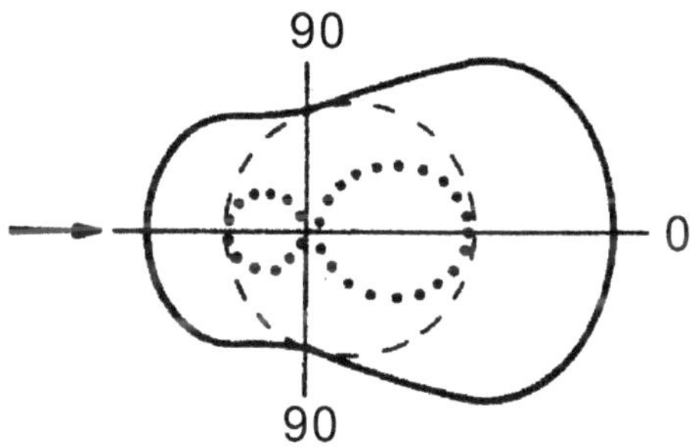

FIGURE 5.6 Unsymmetrical scattering.

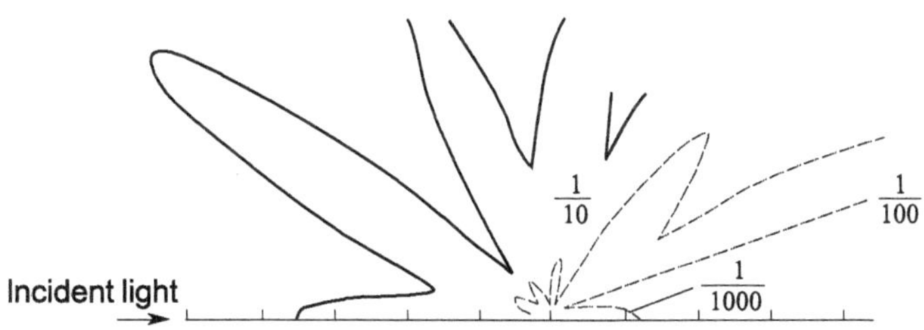

FIGURE 5.7 Distribution of the intensity of dispersed light with the observation angle.

angles is possible and maxima and minima of scattering can be observed at different angles as shown in Fig. 5.7.

5.5 LIGHT SCATTERING OF MACROMOLECULE SOLUTIONS

5.5.1 Fluctuation in Density and Concentration

As discussed previously, Rayleigh assumed that the concentration of colloid is very small; far enough apart, there are no interactions between particles, thus the intensity of scattered light of the system is equal to the sum of the intensity of scattered light of all particles. But in liquid or solution the molecules are close enough together so that interference occurs between the waves emitted from different centers. Therefore, the Rayleigh equation cannot be applied to the light scattering of solution. In this case, the theory of density fluctuation is used to explain it. In this theory, it is not the individual molecules that are the scattering centers, but rather the small domains of compression or rarefaction that arise from fluctuations caused by thermal motion of molecules. In other words, these fluctuations make the density and concentration of the small domain rise up or fall down instantaneously, leading to the rising up or falling down of refractive index of the small domains instantaneously, thus producing scattered light. From this viewpoint, the next relationship was proposed:

$$R_\theta \propto \overline{\Delta c^2} \tag{5.6}$$

where c is the concentration of solution and Δc the fluctuation in concentration. This equation shows the Rayleigh ratio is proportional to the mean square value of fluctuation in concentration. Since the fluctuation is caused by thermal motion, the next relationship was also proposed:

$$\overline{\Delta c^2} \propto K_B T \tag{5.7}$$

where K_B is the Boltzmann constant, and T the thermodynamics temperature. Since the fluctuation in concentration will result in concentration difference, which, in turn, will result in osmotic pressure, which, in turn, will suppress the fluctuation in concentration, the next relation was also proposed:

$$\overline{\Delta c^2} \propto \frac{1}{\left(\frac{\partial \pi}{\partial c}\right)} \tag{5.8}$$

where π is osmotic pressure. In addition, the fluctuation is also related to the size of concentration and small volume element, therefore, we also have

$$\overline{\Delta c^2} \propto c \tag{5.9}$$

and

$$\overline{\Delta c^2} \propto \frac{1}{\Delta V} \tag{5.10}$$

where ΔV is the volume of small volume element in which fluctuation takes place. In summary, we have

$$\overline{\Delta c^2} \propto \frac{K_B T c}{\Delta V\left(\frac{\partial \pi}{\partial c}\right)} \tag{5.11}$$

Accordingly, Rayleigh ratio is given by

$$R_\theta = \frac{i_\theta r^2}{I} = \frac{KcRT}{\left(\frac{\partial \pi}{\partial c}\right)}(1+\cos^2\theta) \tag{5.12}$$

where

$$K = \frac{2\pi^2 n^2}{N_A \lambda_0^4}\left(\frac{dn}{dc}\right)^2 \tag{5.13}$$

It can be seen that Eq. (5.12) is very similar to the Rayleigh equation.

5.5.2 Measurement of the Molecular Weight of Polymers

From Eqs. (4.5) and (4.6), we know the osmotic pressure of the macromolecule solution is given by

$$\pi = cRT\left(\frac{1}{M} + A_2c + \cdots\cdots\right)$$

thus

$$\frac{\partial \pi}{\partial c} = RT\left(\frac{1}{M} + 2A_2c + \cdots\cdots\right)$$

For dilute solution, it can be shortened as follows

$$\frac{\partial \pi}{\partial c} = RT\left(\frac{1}{M} + 2A_2c\right) \tag{5.14}$$

Substituting Eq. (5.14) into Eq. (5.12) and supposing $\theta = 90$ degree yields

$$R_{90} = \frac{Kc}{\left(\frac{1}{M}\right) + 2A_2c} \tag{5.15}$$

$$\frac{Kc}{R_{90}} = \frac{1}{M} + 2A_2c \tag{5.16}$$

By the plot of Kc/R_{90} versus c, we can obtain the intercept and the slope, thus work out the molecular weight and the second virial coefficient A_2. Note that the molecular weight, which is measured by the light scattering method, is weight average molecular weight.

5.6 TURBIDITY OF WATER

If a water sample contains colloidal materials or suspended impurities, it will appear to be not transparent or turbid. The turbid appearance of surface water is caused by mud, sand, clay, and organic materials. Generally speaking, the underground water is transparent, but if it contains ferrous ions, ferric hydroxide will form after contact with the air, leading to a brown-yellow color and turbid appearance. Most industrial wastewater and domestic sewage are turbid because they commonly contain colloidal and/or suspended inorganic and/or organic impurities.

Suppose the thickness of medium is Δx, the intensity of incident beam I, the energy attenuation I_s, and the intensity of transmitted light I_t. Then the intensity of transmitted light is given by

$$I - I_s = I \exp(-\tau \cdot \Delta x) \tag{5.17}$$

thus

$$\frac{I-I_s}{I}=\exp(-\tau\cdot\Delta x)$$

$$\frac{I}{I-I_s}=\exp(\tau\cdot\Delta x)$$

$$\ln\frac{I}{I-I_s}=\tau\cdot\Delta x$$

$$\ln\frac{I}{I_t}=\tau\cdot\Delta x \tag{5.18}$$

or

$$A=\tau\cdot\Delta x \tag{5.19}$$

Eqs. (5.18) and (5.19) are the same as the equations of Lambert's law in form, and we know $\ln(I/I_t)$ or A in the equations of Lambert's law is the absorbance of solution, but here it is obtained from light scattering, not from the light absorbance, therefore, is referred to as pseudo-absorbance. Eq. (5.19) may be written as

$$\tau=\frac{A}{\Delta x} \tag{5.20}$$

where τ is defined as turbidity, which is the attenuation coefficient of the energy of incident beam per unit optical path due to light scattering. In practice, turbidity represents the turbid extent of water.

In the early years, the optical effect produced by insoluble silicon materials, for example, kaolin or discoloring clay, was used as the basis of the turbidity measurement. In this method, firstly, the standard turbid liquid is prepared by using the insoluble silicon materials, and the turbidity of dispersion of 1 mg dm^{-3} SiO_2 is defined as 1 degree. Comparing the turbid extent of a water sample to be measured with that of the standard turbid liquid gives the turbidity of the water sample, regardless of the size and shape of the particles in them; that is, if the turbid extent of a water sample equals to that of the standard turbid dispersions of n mg dm^{-3} SiO_2, the turbidity of this water sample will equal n degree. Currently, the kaolin or discoloring clay is sieved by passing through 200-mesh screen before use, and then is used as the general standard material to prepare the standard turbid dispersions according to the specified procedures. This turbidity unit is called "silicon unit." Another standard turbid dispersion is Formazin polymer solution, which is prepared by mixing hydrazine sulfate and hexamethylenetetramine (urotropine). The preparation of Formazin polymer solution is very simple, and its reproducibility of light-scattering property is better than other standard dispersions. Therefore, Formazin polymer solution is widely used today. The preparation procedures of standard Formazin polymer solution is introduced in detail as follows.

Firstly, dissolve 1.000 g of hydrazine sulfate in distilled water and dilute it using distilled water to 100 cm^3 of volume in a volumetric flask. Similarly, dissolve 10.000 g of hexamethylenetetramine in distilled water and dilute it using distilled water to 100 cm^3 of volume in the other volumetric flask. Then draw 5 cm^3 of the solutions from the aforementioned two volumetric flasks respectively, and mix them in a 100 cm^3 volumetric flask, let it stand and react for 1 h at 25°C, and then dilute it using distilled water to 100 cm^3 of volume after cooling. The turbidity of the solution prepared previously is 400 degree. This solution can be kept for 1 month at common temperatures.

The unit of turbidity measured by light-scattering photometer is defined as NTU, while the unit of turbidity measured using standard Formazin polymer solution is FTU. It is known that 1 FTU = 1 NTU.

When using the turbidity, we should distinguish it from the chromaticity, which is one of the water quality indexes. Some of water may be colorful, thus leading light absorption, but clear and not turbid. In addition, we should know that the turbidity does not equal the suspended substance (SS), which is also one of the water quality indexes. SS is the mass of substance, which can be retained by filter paper, while turbidity is an optical effect that depends not only on the value of SS, but also on the size and shape of particles. If two water samples have the same SS but different particle size distributions, their turbidities are commonly different.

Turbidity is a very important index for drinking water. Although the inorganic matters in water, such as mud and sand, are not directly harmful to human health, there are lots of bacteria adsorbed on them, which will decrease the effect of disinfection. There are certain requirements for turbidity in various industrial waters, for example, for the water supplied to the boilers, the turbidity should be as low as possible to reduce scaling. Many kinds of industries need low turbidity water, such as food processing, dyeing, and paper making. Turbidity is one of the characteristics of water that can be used to decide whether water is polluted or not from its appearance. In summary, the measurement and removal of turbidity are very important tasks in water and wastewater treatment.

Chapter 6

Rheology Properties

Chapter Outline

6.1 BASIC CONCEPT AND BASIC THEORY

Rheology is the science of the deformation and flow of matter. The study of rheology has contributed much toward clarifying ideas concerning the nature of colloidal systems. First we introduce the basic concept and basic theory.

6.1.1 Shearing Strain and Shear Rate

Let us consider a solid cube with surface area A and height y, as shown in Fig. 6.1. Suppose a tangential couple of force F and force F' is applied to the bottom and upper surfaces respectively at the same time. The upper surface will move x distance from the original position and produce an angle θ between the side surface and its original position. We know that if F is constant, the θ will not change. In this instance, F/A is shear stress, which is the force applied to the unit surface and commonly represented by τ (N m^{-2}). The effect of shear stress is to overcome the friction between the parallel layers of the cube, producing the relative displacement of layers; as a result, the shape change of the cube will

Colloid and Interface Chemistry for Water Quality Control. http://dx.doi.org/10.1016/B978-0-12-809315-3.00006-2

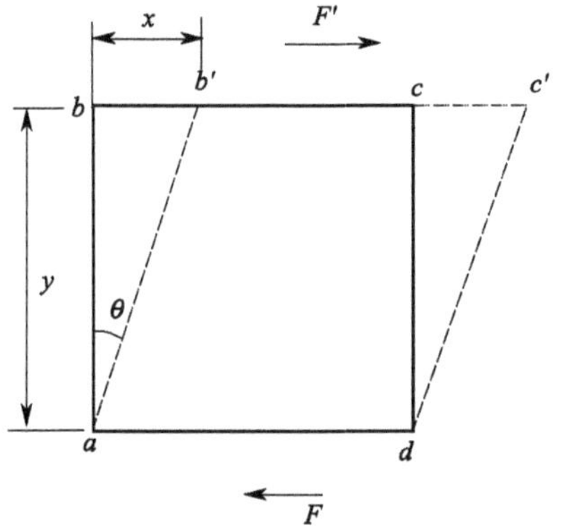

FIGURE 6.1 **Profile map of the shearing strain produced by shear stress.**

take place. In addition, θ is referred to as the shearing strain and equals x/y. According to Hooke's law, in elastic range $\theta = (1/G)\tau$, where G is the stiffness coefficient, which shows the capacity of solid to resist shape change.

If a tangential couple of force F and force F' is applied to a liquid cube, the situation will be different from that discussed previously. Although the shear stress remains unchanged, θ will constantly increase with time, that is, flow will take place, thus producing shear rate $d\theta/dt$ which is represented by $D(s^{-1})$:

$$D = \frac{d\theta}{dt} = \frac{d(\frac{x}{y})}{dt} = \frac{1}{y} \cdot \frac{dx}{dt} = \frac{v}{y} \tag{6.1}$$

Eq. (6.1) shows D is the velocity gradient in the vertical direction. For non-laminar flow

$$D = \frac{dv}{dy} \tag{6.2}$$

6.1.2 Newton Equation

For pure liquid or the solution of small molecules in laminar flow, the shear stress is proportional to the velocity gradient

$$\tau = \eta D \tag{6.3}$$

When multiplied by area A, Eq. (6.3) becomes

$$F = \eta A \frac{dv}{dy} \tag{6.4}$$

Eqs. (6.3) and (6.4) are Newton equations where η is the coefficient of viscosity, which is a measure of the internal friction offered to the relative motion of different parts of the liquid. The coefficient of viscosity (or viscosity, for short) of a fluid has dimensions kilogram-meter per second (abbreviated as kg m^{-1} s^{-1}) in SI units, which is called a pascal-second (abbreviated as Pa s). We know that the cgs unit of viscosity, the poise (abbreviated as *P*), is widely used and that 10 *P* = 1 kg m^{-1} s^{-1}, that is, 10 *P* = 1 Pa s. Fluids that obey the form predicted by Eqs. (6.3) and (6.4) are referred to as Newtonian. For a Newtonian fluid, η is a well-defined quantity for a given temperature and pressure, and independent of τ and *dv/dx*, provided that the flow is laminar. Fluids that do not obey the form predicted by Eqs. (6.3) and (6.4) are referred to as non-Newtonian. For a non-Newtonian fluid, η is a function of the shear rate (or the velocity gradient).

6.2 MEASUREMENT OF VISCOSITY

6.2.1 Capillary Flow Method

Suppose we have a capillary tube through which fluid flows under pressure *P*, as shown in Fig. 6.2.

Let the length of the capillary tube be *L* and the distance from the wall of the tube to the center of the fluid be *y*, whereas the distance from the center of the fluid to the wall of the tube is *r* and the radius of the tube is *R*. Assuming all of the force driving the liquid to flow is used to overcome the viscosity resistance, we will have

$$\pi r^2 P - 2\pi r L\eta \frac{dv}{dy} = 0$$

$$\pi r^2 P + 2\pi r L\eta \frac{dv}{-dy} = 0$$

$$\pi r^2 P + 2\pi r L\eta \frac{dv}{dr} = 0$$

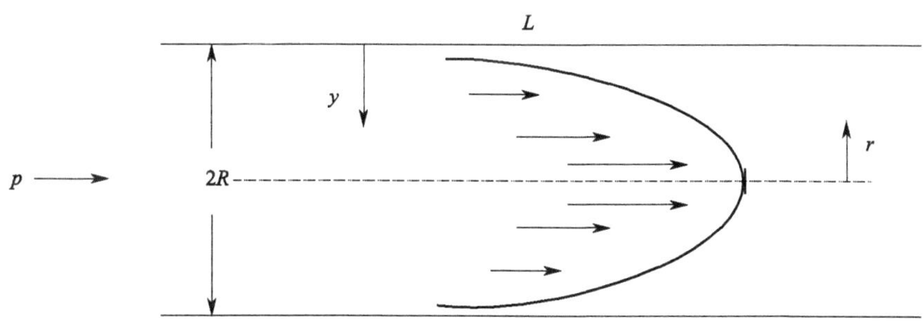

FIGURE 6.2 **Distribution of flow rate and velocity gradient of fluid in a capillary tube.**

Thus

$$\frac{dv}{dr} = -\frac{1}{\eta}\frac{Pr}{2L} \tag{6.5}$$

Eq. (6.5) indicates that the velocity gradient is asymmetrical and changes with the distance from the center of the fluid to the wall of the tube. At the center, since $r = 0$, $dv/dr = 0$; that is, the velocity gradient reaches the minimum. At the wall, since $r = R$, $(dv/dr) = -(1/\eta)(PR/2L)$; that is, the velocity gradient reaches the maximum. According to Eq. (6.5)

$$dv = -\frac{1}{\eta}\frac{Pr}{2L}dr$$

Integrating it from the wall to r gives

$$\int_0^{v(r)} dv = \int_R^r -\frac{1}{\eta}\frac{Pr}{2L}dr$$

$$v(r) = -\frac{1}{\eta}\frac{P}{2L}\int_R^r r\,dr = -\frac{P}{4\eta L}(r^2 - R^2) = \frac{P}{4\eta L}(R^2 - r^2)$$

Since the perimeter of the layer of liquid at r is $2\pi r$ and the thickness dr, the section area of this layer will be $2\pi r{\cdot}dr$, thus the volume of outflow liquid per second from the tube equals the sum of that from all of the layers

$$Q = \int_0^R 2\pi r v(r)dr = \int_0^R 2\pi r \frac{P}{4\eta L}(R^2 - r^2)dr = \frac{\pi P}{2\eta L}\int_0^R r(R^2 - r^2)dr = \frac{\pi PR^4}{8\eta L} \tag{6.6}$$

thus

$$\eta = \frac{\pi PR^4}{8QL} = \frac{\pi PR^4 t}{8VL} \tag{6.7}$$

Eq. (6.7) is a Poiseuille equation, in which t is the time needed for the liquid of volume V to outflow, and V is the volume of liquid that flows out in time t. Substituting the volume V and the time t measured in Eq. (6.7) will give the viscosity of liquid. This method is referred to as an absolute method.

Actually, not all of the force P driving the liquid to flow is used to overcome the viscosity resistance; some of it is consumed to produce heat. In addition, the streamline at the end of the tube is different from that inside the tube. Therefore, we have to correct P for the kinetic energy and correct L for the end. The former is to substitute $(P - \Delta P)$ for P, the latter is to substitute $(L + nR)$ for L. After the corrections, Eq. (6.7) becomes

$$\eta = \frac{\pi R^4 Pt}{8V(L+nR)} \quad m\frac{\rho V}{8\pi(L+nR)t} \tag{6.8}$$

where m is instrument constant, and ρ is the density of liquid. If the capillary tube is vertical, P will be the static pressure of the liquid column as follows

$$P = \bar{h}\rho g \tag{6.9}$$

where $\bar{h}$ is the average height of the liquid during the flow process. Substituting it into Eq. (6.8) and combining the constants gives

$$\eta = A\rho t - B\frac{\rho}{t} \tag{6.10}$$

If we know the densities and outflowing time of two liquids, the constants A and B can be calculated by solving the equation group. And then the viscosity of any other liquid can be obtained by this capillary tube through the experiment. If the capillary tube is so long that the outflowing time exceeds 100 s, the second term of Eq. (6.10) can be omitted, thus we have

$$\eta \approx A\rho t \tag{6.11}$$

This equation is convenient for determining the viscosity of liquid (see latter).

The most frequently employed methods for measuring viscosity is based on the flow through a capillary tube as discussed previously. There are two common viscometers. One is Ostwald and the other is Ubbelohde, as shown in Fig. 6.3. When it is used, sufficient liquid is added to the viscometer through

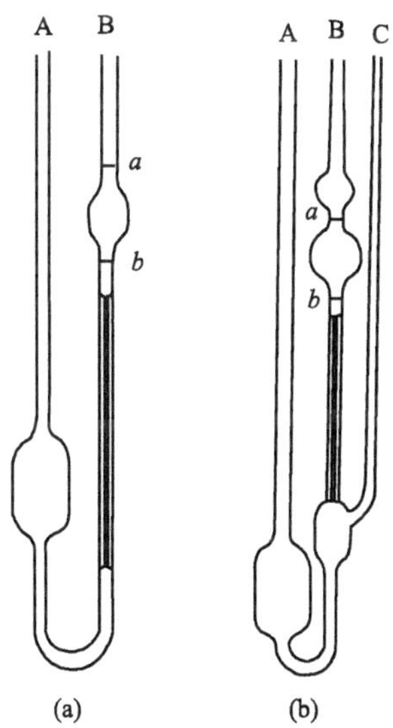

FIGURE 6.3 Capillary viscometers. (a) Ostwald viscometer and (b) Ubbelohde viscometer.

tube A, and the liquid is drawn up into tube B until the liquid level is above mark A. The liquid is then released and the time for the meniscus to pass the volume between marks A and B is measured. Substituting this flow time into Eq. (6.10) gives the viscosity of liquid.

In the Ostwald viscometer, the liquid in the capillary tube is directly connected with the liquid in the left bulb; therefore, the flow time depends on the liquid volume in the apparatus, that is, relates to the pressure difference between tube A and B. If you want to measure the viscosity of liquids with different concentrations, you can not dilute the solution in the viscometer because it will increase the volume of liquid. Before measuring, you must discharge the original liquid, and add new liquid with different concentrations and the same volume after washing and drying the apparatus. When using the Ubbelohde viscometer, you should close tube C before the liquid is drawn up into tube B. After the liquid is drawn up and before releasing it, open tube C so that the liquid in the bottom bulb connecting with the capillary tube fall down automatically. As a result, the liquid in the capillary tube will become disconnected with the liquid in the bulb. In such a case, the flow time will not depend on the liquid volume in the apparatus or the pressure difference between tube A and B, and there will be no need to discharge the original solution, but you need to repeat the measuring procedures after diluting the solution to measure the viscosity of the solution with a different concentration.

6.2.2 Concentric Rotational Cylinder Method

The concentric cylinder viscometer consists of two concentric rotational cylinders. One is the inner cylinder, and the other is the outer cylinder. As shown in Fig. 6.4, the inner cylinder is connected with a light-reflecting mirror and suspended by a torsion wire.

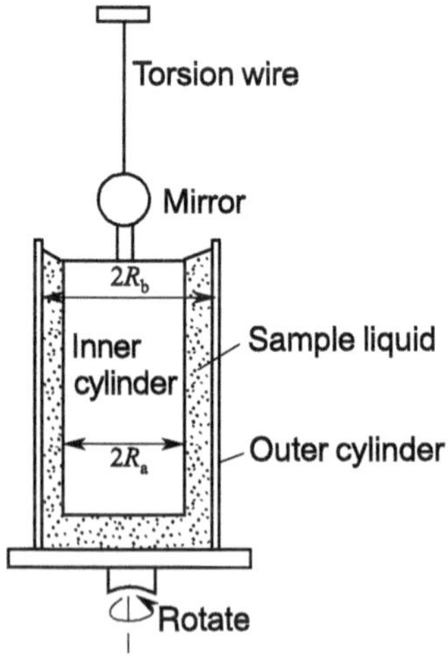

FIGURE 6.4 Concentric cylinder viscometer.

First, add the liquid into the gap between the outer cylinder and the inner cylinder, then make the outer cylinder rotate at a constant rate. The inner cylinder will rotate with the outer cylinder due to the viscosity of liquid in the gap, but the torsion wire will make it rotate back at the same time. When the two opposite torques are equal, the inner cylinder will stop rotating. The torsion angle θ can be measured by the reflecting light from the mirror. Let the height of the inner cylinder be L; the radius of the inner cylinder and outer cylinder be R_a and R_b respectively; the rotating rate of the outer cylinder be ω; and the viscosity of the liquid be η, for Newtonian fluid; the torque at equilibrium is

$$T = 4\pi L \frac{R_a^2 \cdot R_b^2}{R_b^2 - R_a^2} \eta\omega = K\eta\omega \tag{6.12}$$

or

$$T = K'\theta \tag{6.13}$$

where K and K' are the instrument constants. For two kinds of liquids

$$\frac{T_1}{T_2} = \frac{K\eta_1\omega_1}{K\eta_2\omega_2}$$

and

$$\frac{T_1}{T_2} = \frac{K'\theta_1}{K'\theta_2}$$

thus

$$\frac{\theta_1}{\theta_2} = \frac{\eta_1\omega_1}{\eta_2\omega_2} \tag{6.14}$$

According to Eq. (6.14), if we know the viscosity of one of the two liquids, and also determine the rotating rates and torsion angles of the two liquids, respectively, by experiment, we can work out the viscosity of the other liquid.

The concentric rotational cylinder method is suitable for liquid of high viscosity.

6.3 VISCOSITY OF DILUTE COLLOIDAL SOLUTIONS

6.3.1 Basic Concept

When colloidal particles are dispersed in a fluid, the flow of the fluid is disturbed and the viscosity becomes higher than that of the pure fluid. The increment of viscosity is of greater significance than absolute viscosity, therefore, the following functions of viscosity are defined.

Relative viscosity:

$$\eta_r = \frac{\eta}{\eta_0} \tag{6.15}$$

Specific viscosity:

$$\eta_{sp} = \frac{\eta - \eta_0}{\eta_0} = \eta_r - 1 \tag{6.16}$$

Reduced viscosity:

$$\frac{\eta_{sp}}{c} \tag{6.17}$$

Intrinsic viscosity:

$$[\eta] = \lim_{c \to 0} \frac{\eta_{sp}}{c} \tag{6.18}$$

where η_0 is the viscosity of pure fluid, η the viscosity of dispersion, and c the concentration of particles; intrinsic viscosity equals the limiting value of the reduced viscosity when the concentration approaches zero. Since the interaction between particles can be omitted when the concentration approaches zero, intrinsic viscosity reflects only the friction between particles and medium.

6.3.2 Effect of Spherical Particles on the Viscosity of Colloidal Dispersion

If there is a spherical particle in fluid, the fluid will surround the particle to flow forward, thus forming a longer streaming line, leading to greater friction, as shown in Fig. 6.5.

Einstein obtained a famous equation for the dilute dispersion containing rigid spherical particles as follows

$$\eta = \eta_0 (1 + 2.5\phi) \tag{6.19}$$

where ϕ is the volume fraction of the particles (ie, the ratio of the particle volume to the total volume of dispersion). It can be seen that the effects of the

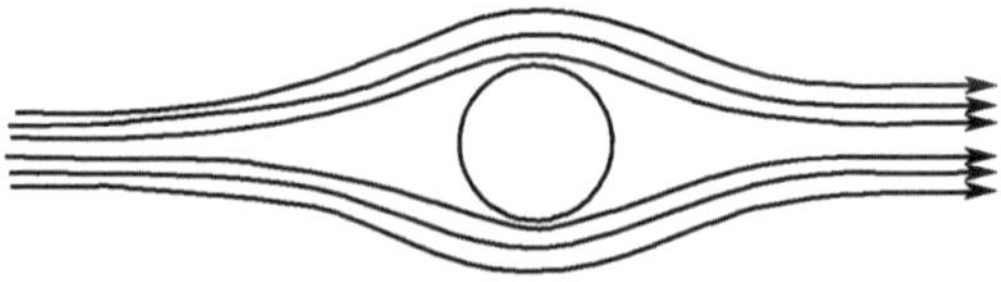

FIGURE 6.5 Effect of spherical particles on viscosity of fluid.

particles on the viscosity of a dispersion depends only on the total volume that they occupy and is independent of their particle size. Therefore, we cannot know the size of individual particles from this equation.

6.3.3 Effect of Particle Morphology on the Viscosity of Colloidal Solutions

Particle asymmetry has a marked effect on the viscosity of fluid. There are two cases:

1. The viscosity is higher than that predicted by the Einstein equation and increases with the increase in axial ratios. There are three reasons for such cases: The first one is that the streaming line is disturbed much stronger than spherical particles; the second one is that the particles turn in the velocity gradient field to increase the disturbance to the flow field; the third one is that the particles disturb one another to increase the resistance to flow.
2. Velocity gradient makes the particle oriented along the streaming line, which will decrease the internal friction, as shown in Fig. 6.6. In such cases, η in Eq. (6.3) is not still a constant, but decreases with the increase in velocity gradient.

6.3.4 Effect of Particle Solvation on the Viscosity of Colloidal Solutions

The volume fraction term ϕ in viscosity equations includes any solvent that acts kinetically as a part of the particles; that is, the solvation will increase the volume fraction of the particles, therefore increasing the viscosity of the fluid. For spherical particles

$$\phi_{\text{solvation}} = \left(1 + \frac{3\Delta R}{R}\right)\phi_{\text{dry}} = K'\phi_{\text{dry}} \tag{6.20}$$

where ΔR is the thickness of the solvent layer, and $K' > 1$.

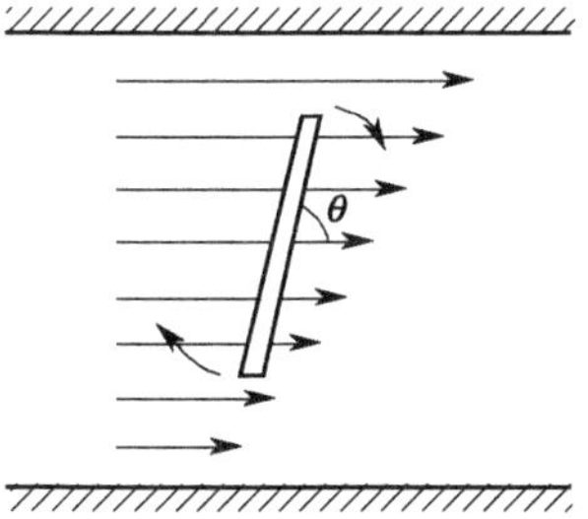

FIGURE 6.6 Orientation of particles in velocity gradient field.

6.3.5 Measuring the Molecular Weight of Polymers and Flocculants in Water Treatment

For most linear polymers in solution the chains are somewhat more extended than random, and the relationship between intrinsic viscosity and molecular weight is expressed by the Mark–Houwink equation:

$$[\eta] = KM^{\alpha} \tag{6.21}$$

where M is molecular weight, and K and α are constants characteristic of the polymer-solvent system (α depends on the configuration of polymer chains) and approximately independent of molecular weight. The values of α are defined as follows:

1. Spherical particle: $\alpha = 0$
2. Rigid rod molecule: $\alpha = 2$
3. Threadlike macromolecule: $\alpha = 0.5 \sim 1.0$. In a poorer solvent, the α value is smaller because the molecule chain shrinks to be a coil, but in a good solvent, the α value is greater because the molecule chain is more extended.

Example 6.1 Variation of Viscosity with Polymer Configuration

Suppose the molecular weight of cellulose triacetate is 10^5, and polyisobutylene has the same molecular weight as the cellulose triacetate. For the cellulose triacetate in acetone at 25°C, $K = 8.97 \times 10^{-5}$ dm^3 g^{-1}, $\alpha = 0.9$, and for polyisobutylene in benzene at 25°C, $K = 1.07 \times 10^{-3}$ dm^3 g^{-1}, $\alpha = 0.5$. Calculate the viscosities of these two systems, respectively, and explain the relationship between these values and the system property.

Solution: Intrinsic viscosities are calculated by directly substituting the aforementioned data into Eq. (6.21).

For cellulose triacetate:

$$[\eta] = KM^{\alpha} = 8.97 \times 10^{-5} \times (1 \times 10^5)^{0.9} = 2.84(\text{dm}^3\,\text{g}^{-1})$$

For polyisobutylene:

$$[\eta] = KM^{\alpha} = 1.07 \times 10^{-3} \times (1 \times 10^5)^{0.5} = 0.34(\text{dm}^3\text{g}^{-1})$$

This result indicates that the cellulose triacetate has an intrinsic viscosity of more than eight times greater than the polyisobutylene, even though the two polymers have the same molecular weight. These results mainly depend on α, which reflects the configuration of the polymer molecule in solution. Cellulose triacetate repeat units are a six-member ring that carry three bulky acetate groups each. Internally, the rings are inflexible with respect to rotation and because of the bulk of the acetate groups, are expected to be severely hindered in their rotation with respect to each other. As a consequence, such a molecule exists in solution in a highly extended form that approaches the predictions of the free-draining model in intrinsic viscosity. Polyisobutylene, by contrast, is an aliphatic hydrocarbon for which low molecular weight aliphatic molecules are expected to be better solvents. Although benzene dissolves this polymer, it is a poorer solvent because of the polarizable π electrons. It is therefore plausible that this system corresponds to theta conditions with an α value of 0.5.

If you want to determine the molecular weight of a polymer by Eq. (6.21), firstly, you should determine the intrinsic viscosity using Huggins empirical equations, which is shown as follows.

$$\frac{\eta_{sp}}{c} = [\eta] + K'[\eta]^2 c \tag{6.22}$$

$$\frac{\ln \eta_r}{c} = [\eta] - \beta[\eta]^2 c \tag{6.23}$$

It can be seen from Eq. (6.22) that a plot of η_{sp}/c versus c should be a straight line, and from the intercept of which we can obtain the intrinsic viscosity. Similarly, it can be seen from Eq. (6.23) that a plot of $\ln \eta_r/c$ versus c is also a straight line, and from the intercept of which we can also obtain the intrinsic viscosity. Generally, it is requested to plot both of the straight lines respectively in one coordinate system. If the two lines intersect at the same one point of y-axis, the result will be accurate and reliable.

Example 6.2 Determination of the Molecular Weight of a Polymer by Measuring Viscosity

The outflow times of the solution of polyvinyl alcohol of different concentrations are measured at 25°C by Ubbelohde viscometer as shown in the following table.

Concentration (g 100 cm^{-3})	0	0.219	0.291	0.445	0.602	0.704	0.844
Flow time (s)	103.2	114.8	119.2	128.2	138.0	145.4	155.8

Suppose $K = 2.0 \times 10^{-4}$ (100 cm^3 g^{-1}), $\alpha = 0.76$, $\eta_0 = 1.0 \times 10^{-3}$ Pa s. Determine the molecular weight of the polyvinyl alcohol.

Solution

Since the concentration differences of the different solutions are small, we can assume the values of ρ is the same for all solutions, therefore, $\eta_r = \frac{\eta}{\eta_0} \approx \frac{t}{t_0}$ according to Eq. (6.15) and Eq. (6.11). By using Eqs. (6.15)~(6.17) and the aforementioned data, the different functions of viscosity are calculated as shown in following table:

Concentration (g 100 cm^{-3})	0	0.219	0.291	0.445	0.602	0.704	0.844
Flow time (s)	103.2	114.8	119.2	128.2	138.0	145.4	155.8
η_r	—	1.112	1.155	1.242	1.337	1.409	1.510
$\ln \eta_r$	—	0.106	0.144	0.217	0.290	0.343	0.412
η_{sp}	—	0.112	0.155	0.242	0.337	0.409	0.510
η_{sp}/c	—	0.511	0.533	0.544	0.560	0.581	0.604
$\ln \eta_r/c$	—	0.484	0.495	0.488	0.481	0.487	0.488

Plotting the straight line of η_{sp}/c versus c and $\ln \eta_r/c$ versus c, respectively, gives

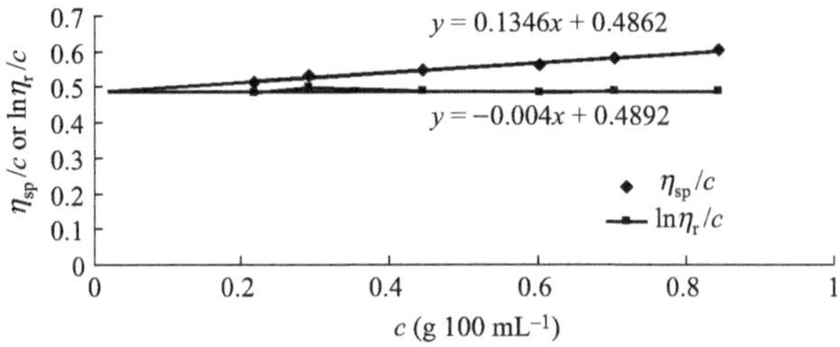

Thus we know the intercept equals 0.49, that is, $[\eta] = 0.49$. Therefore

$$[\eta] = kM^{\alpha}$$

$$0.49 = 2.0 \times 10^{-4} M^{0.76}$$

$$M = 28802$$

Polyacrylamide (PAM) is widely used as the flocculant in water treatment, which can be grouped according to their molecular weight as follows:

1. Low molecular weight: $1 \times 10^3 \sim 1 \times 10^5$
2. Medium molecular weight: $1 \times 10^5 \sim 1 \times 10^6$
3. High molecular weight: $1 \times 10^6 \sim 5 \times 10^6$
4. Very high molecular weight: $>5 \times 10^6$

Currently, the highest molecular weight of PAM has reached up to 1.8×10^7, but the molecular weight of PAM suitable for water treatment is in the range of $4 \times 10^6 \sim 6 \times 10^6$. Generally, the molecular weight of PAM is measured by capillary viscometer. The relations between the intrinsic viscosity and molecular weight are given by the Mark–Houwink equation as follows.

Weight average molecular weight:

$$[\eta] = 3.73 \times 10^{-4} \overline{M}_w^{\,0.66} \text{ (solvent: 1 mol dm}^{-3}\text{ NaNO}_3\text{, 30°C)} \tag{6.24}$$

Number average molecular weight:

$$[\eta] = 6.8 \times 10^{-4} \overline{M}_n^{\,0.66} \text{ (solvent: water, 25°C)} \tag{6.25}$$

Z average molecular weight:

$$[\eta] = 6.31 \times 10^{-5} \overline{M}_z^{\,0.80} \text{ (solvent: water, 25°C} \tag{6.26}$$

where 100 $cm^3\ g^{-1}$ is used as the unit of intrinsic viscosity for all of the aforementioned equations. For the hydrolyzed PAM, the solution of $NaNO_3$ must be used as the solvent to eliminate the electroviscous effect.

6.4 RHEOLOGY PROPERTIES OF CONCENTRATED DISPERSION SYSTEMS

There are many concentrated dispersion systems produced in various productions, for example, the sludge in the settling pond of water plants and the activated sludge of sewage plants. Their rheology properties are very complicated, and very important for practice usage. Plotting the lines of shear rate D versus shear stress τ gives the rheology curve, as shown in Fig. 6.7.

Newtonian fluid and the four kinds of non-Newtonian fluids are discussed as follows.

1. *Newtonian fluid*: As described in the section Newton Equation, fluids that obey the form predicted by Eqs. (6.3) and (6.4) are referred to as Newtonian. For Newtonian fluid, the relationship between D and τ is represented by a straight line of zero intercept, which means that the liquid can flow even at any small shear stress. It is enough for us to use viscosity only to characterize the property of Newtonian fluid. It can be also seen from the $D-\tau$ straight line that the less the slope, the greater the viscosity. Most pure liquids are Newtonian fluids, such as water, glycerol, oil of low viscosity, solution of substance of low molecular weight, dilute colloid dispersion, etc.
2. *Plastic fluid*: Plastic fluid is also called Bingham plastics. Plastic fluid does not flow noticeably until the shearing stress exceeds a certain minimum value, which is termed the yield value. The rheology curve of plastic fluid is also a straight line, which does not go through the original point of the coordinate system, but intersects with shear stress axis at τ_y (yield value), as shown in Fig. 6.7. When we squeeze toothpaste, if the force used is too little, the toothpaste will not flow out of the tube but only become convex and then retract after loosening our grip; if the force used is greater to some extent, the toothpaste will flow out of the tube and never retract. Therefore, toothpaste is considered to be plastic fluid.

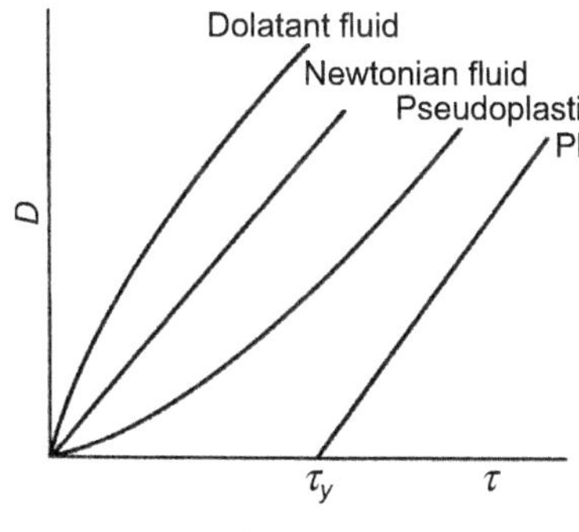

FIGURE 6.7 **Four kinds of non-Newtonian fluids.**

The straight line section of the rheology curve of plastic fluid is represented by

$$\tau - \tau_y = \eta_p D (\tau > \tau_y) \tag{6.27}$$

where η_p is plastic viscosity or structural viscosity. Both η_p and τ_y are very important parameters of plastic fluid. The explanation for plastic fluid is given as follows.

When the concentration of dispersion is too high to result in the contact of particles, the three-dimensional structure will form, as shown in Fig. 6.8. In this case, the value of τ_y represents the structure strength. When shearing stress exceeds $\tau_{y,}$ the structure will be destroyed, leading to flow of the system. Therefore τ_y is the additional necessary force to make the system start to flow. Because the formation and destruction of system structure takes place at the same time, and will reach equilibrium, the total extent of structure destroy is kept constant, and the system will have a nearly invariable viscosity η_p.

3. *Pseudoplastic fluid*: Pseudoplastic fluid is characterized by a gradual decrease in apparent viscosity η_a with increasing rate of shear. This means that the more quickly you stir, the more dilute the system will appear to be. The rheology curve of pseudoplastic fluid goes through the original point of coordinate system, thus has no yield value, and is concave to the axis of shearing stress, as shown in Fig. 6.7.

Pseudoplastic fluid is a common non-Newtonian fluid, such as most macromolecule solutions, emulsions, etc. For this kind of fluid, the relation of $D-\tau$ is given by

$$\tau = KD^n (0 < n < 1) \tag{6.28}$$

where both K and n are the experience coefficients, which depend on the properties of liquid. K represents the viscosity of liquid; the greater the value of K, the more viscous the liquid. The value of n is less than 1 and indicates the extent of deviation from Newtonian fluid. The higher the difference between the value of n and 1, the more significant the behavior of non-Newtonian fluid. It can be seen

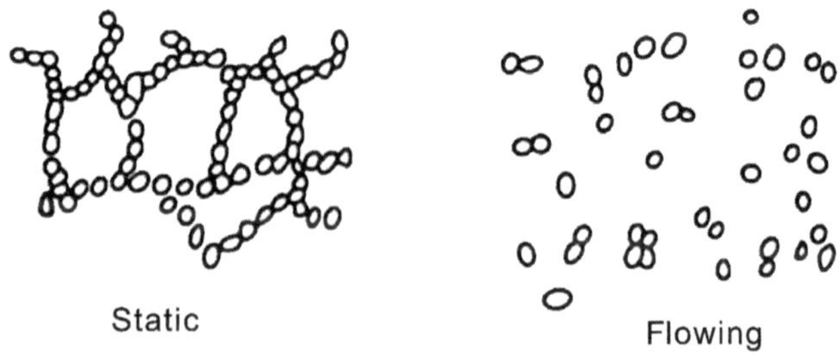

FIGURE 6.8 Explanation for plastic fluid.

that the plot of $\ln\tau$ versus $\ln D$ should be a straight line, and from the intercept and slop of which, we can obtain K and n.

The explanations for pseudoplastic fluids are given as follows.

a. Even though this system has a structure, the structure must be very weak and not easy to recover after destroyed so that τ_y nearly equals zero. Therefore, the apparent viscosity decreases with the increase in shearing rate.
b. Maybe this system has no structure but contains asymmetric particles. Asymmetric particles disturb the flow lines to a greater extent when they are randomly oriented at low velocity gradients than when they have been aligned at high velocity gradients; that is, the decrease of apparent viscosity is caused by the orientation of particles at high velocity gradients.

4. *Dilatant fluid*: Dilatant fluid is characterized by a gradual increase in apparent viscosity η_a with increasing rate of shear. This means that the quicker you stir, the more viscous the system. The rheology curve of dilatant fluid also goes through the original point of the coordinate system, thus has no yield value, and is protruding to the axis of rate of shear, as shown in Fig. 6.7. For this kind of fluid, the relation of $D-\tau$ is also represented by Eq. (6.28), in which $n > 1$.

Two conditions must be met for a dispersion system to become a dilatant fluid: (1) the concentration of the dispersed phase must be high and in a very narrow range; for example, the 40~50% dispersion of starch shows the significant properties of dilatant fluid; and (2) the particles must be dispersed, not aggregated. In the case of smaller shearing stress, all of the particles are dispersed, thus the viscosity is smaller, but in the case of greater shearing stress, the particles contact each other and aggregate to increase the resistance to flow, thus leading to a higher viscosity, as shown in Fig. 6.9. When the concentration is smaller, since the structure does not form easily, dilatant phenomenon will not take place; it will become Newtonian fluid or pseudoplastic fluid. When the concentration is high to form the structure before shearing, the dilatant phenomenon also will not take place.

5. *Thixotropic fluid*: Here we discuss the fluids in which the shear stress/shear rate behavior depends on time, that is, the kinematic history of the fluid.

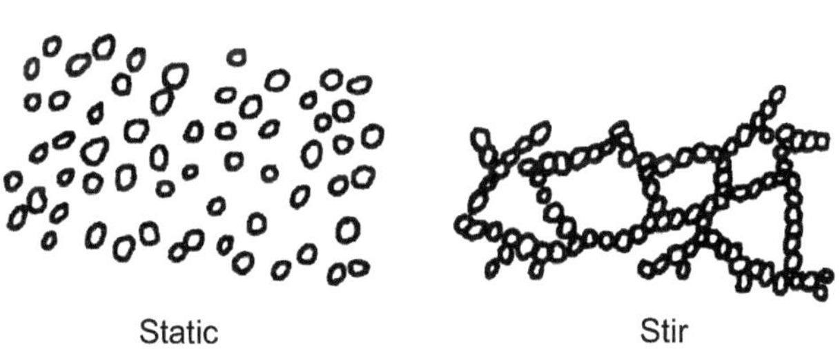

FIGURE 6.9 **Explanation for dilatant fluid.**

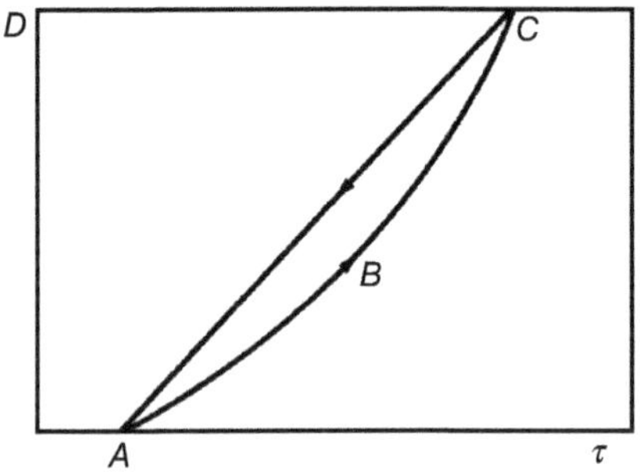

FIGURE 6.10 **Hysteresis loop of thixotropic fluid.**

Suppose we add a concentrated aqueous solution of $Fe_2O_3{\cdot}xH_2O$ or V_2O_5 in a test tube and allow it to stand for some time; it will become semisolid and cannot flow out of the tube. But if a strong shake is applied to the test tube containing the semisolid, it will return to the original flow state. It is known that such changes can be continually repeated. We call this kind of dispersion thixotropic fluid.

There are many explanations for thixotropy. The most popular one follows. Generally speaking, needlelike particles and schistous particles exhibit thixotropy more easily than spherical particles. When the system is allowed to stand, their edges and ends attract each other so that the structure forms. But when followed by a strong shaking, the structure is then broken down by the shear stress. In turn, when left to stand, Brownian motion will restore the system to its original condition. Since Brownian motion is slow, a long time is needed to reform the structure. The rheology curve of thixotropic fluid is similar to that of plastic fluid, except that it is more complicated. A thixotropic hysteresis loop can be obtained by measuring the nonequilibrium shear stress as the shear rate is first increased and then decreased in a standard way, as shown in Fig. 6.10.

6.5 RHEOLOGY PROPERTIES OF SLUDGE PRODUCED IN WATER TREATMENT

It is well known that sludge conditioning and dewatering are very important and necessary operation units both in water treatment plants and sewage treatment work. The previous investigations showed that the flocs sludge produced in water plants, the activated sludge produced in sewage treatment work, and the digested sludge are non-Newtonian fluids. For example, the sludge of some water plant has shear-thinning property and can be simulated by Bingham plastics model. The rheology properties of the non-Newtonian fluids stated previosly depend not only on the surface properties of sludge but also on the composition of sludge, therefore, are very important indications and predictions for the study of sludge conditioning, dewatering, and macro operation. For example, we can determine the flocculant dosage by using viscosity as an index for sludge conditioning in water treatment plants. It was proven that the limiting viscosity and the reduced area of hysteresis loop are related to the concentration of activated sludge and can be used to reflect the expansion of filamentous bacteria. It was also discovered that the shear stress of sludge will change significantly after aerobic digestion or anaerobic digestion.

Chapter 7

Electrical Properties

Chapter Outline

Colloid and Interface Chemistry for Water Quality Control. http://dx.doi.org/10.1016/B978-0-12-809315-3.00007-4

7.1 ORIGIN OF THE SURFACE CHARGE OF COLLOIDS IN NATURAL WATERS

It is well known that colloidal particles have stability with respect to aggregation, and the stability of colloidal particles is mainly due to their surface charge. Generally speaking, the surface charges of dispersed particles are caused by the nonequilibrium distribution of charged ions between particles and solution. The reasons for the nonequilibrium distribution are different depending on the various properties of the particles. Commonly, a surface immersed in natural waters acquires charges in a number of ways, which are discussed in the following sections.

7.1.1 Isomorphous Replacement Within the Lattice

Most inorganic suspended solids and colloidal particles in natural waters come from clay minerals in soil. They are crystals consisting of aluminum or magnesium silicate, in which SiO_2 tetrahedral layers and AlO(OH) octahedral layers are linked to each other by oxygen atoms, forming the plate-layer structure. Fig. 7.1 shows the double layer structure of kaolin, which consists of one tetrahedral layer and one octahedral layer. It is very easy for the OH groups in octahedral layers to form hydrogen bonds with the O atoms of tetrahedral layers in the other double layer structure, thus forming a multilayer plate.

The surface charge at the phase boundary of clay mineral particles may be caused by lattice imperfections at the solid surface and by isomorphous replacements within the lattice. For example, if in any array of solid SiO_2 tetrahedra, an Si atom is replaced by an Al atom (Al has one electron less than Si), a negatively charged framework is established. Similarly, isomorphous replacement of the Al atom by Mg atom in networks of aluminum oxide octahedral also leads to a negatively charged lattice.

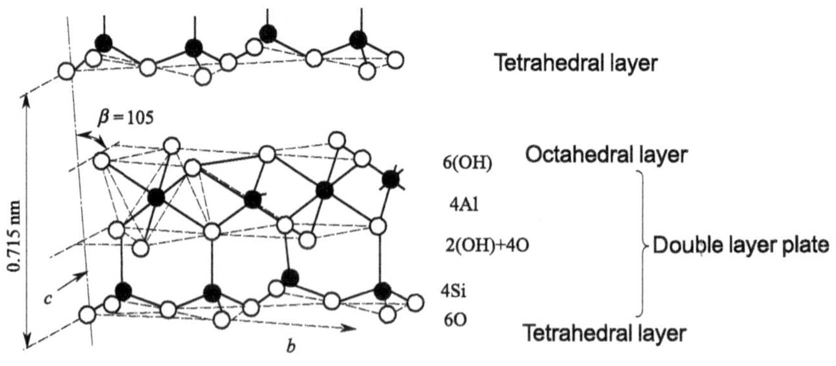

FIGURE 7.1 Structure of kaolin.

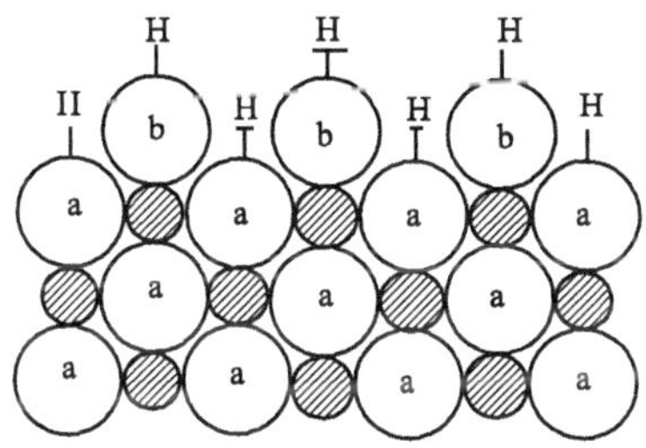

FIGURE 7.2 **Structure of the hydrous oxide mineral. ⓐ Original oxygen atom; ⓑ oxygen atom of hydroxyl group.**

7.1.2 Ionization and Adsorption of Hydrous Oxide Minerals

Essentially, the clay minerals in natural waters are oxides of Si, Fe, Al, etc., hence also called oxide minerals. Commonly, there is one layer of hydroxyls at the surface of the oxide mineral due to hydration. There are two reasons for the hydroxyl formation at the surface of hydrous oxide minerals as follows: (1) H_2O molecules coordinate with Si, Fe, Al, etc., more easily because the coordination numbers of the atoms at the surface are not saturated. These coordinating H_2O molecules then ionize to form hydroxyls. (2) The oxygen atoms at the surface can adsorb the H^+ ions from the solution to form hydroxyls easily because their valences are also not saturated. The structure of the hydrous oxide mineral is shown in Fig. 7.2.

The hydrous oxide minerals acquire their charges mainly through the ionization and adsorption of hydroxyl groups, for example, the ionization and adsorption of silicon hydroxyl groups at the surface is shown in Fig. 7.3.

The ionization and adsorption of these groups, and the net charges, depend strongly on the pH value of the solution. At low pH value the surface will be positively charged and at high pH value it will be negatively charged. The pH value at which the net charge is zero is referred to as the isoelectric point (pzc), and denoted by pH_0 or pH_{pzc}. Most hydrous oxide minerals have a pH_0 lower than that of natural water, for example, the pH_0 of the silicon hydroxyl surface approximately equals 2. Generally, the pH value of natural water is higher than the pH_0 of hydrous oxide mineral, therefore, the clay mineral in natural water is always negatively charged. The produced charge density is given by

$$\sigma_0 = F(\Gamma_{H^+} - \Gamma_{OH^-}) \quad (c\,cm^{-2}) \tag{7.1}$$

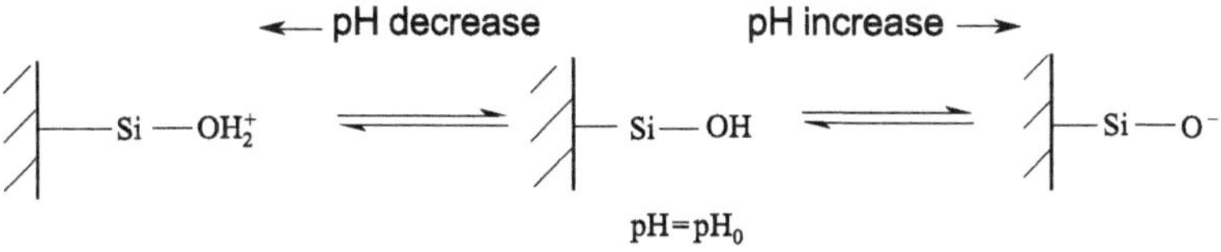

FIGURE 7.3 **Ionization and adsorption of Si-OH groups at the surface.**

where Γ_{H^+} represents the number of H^+ ions adsorbed at the surface of the unit area (mol cm^{-2}); Γ_{OH^-}, the number of OH^- ions adsorbed at the surface of the unit area (mol cm^{-2}); and F, Faraday's constant (96,485 C mol^{-1}).

7.1.3 Specific Adsorption

When water is polluted, the surface charges will become diversified. Charge can also originate by the processes in which solutes become coordinatively bound to solid surfaces, for example,

$$S + S^{-2} = S_2^{2-} \tag{7.2}$$

$$Cu + 2H_2S = Cu(SH)_2^{2-} + 2H^+ \tag{7.3}$$

$$FeOOH + HPO_4^{2-} = FeOHPO_4^- + OH^- \tag{7.4}$$

$$R(COOH)_n + mCa^{2+} = R - [(COO)_n Ca_m]^{2m-n} + nH^+ \tag{7.5}$$

$$MnO \cdot H_2O + Zn^{2+} = MnOOHOZn^+ + H^+ \tag{7.6}$$

The alteration in the surface results from chemical reactions; these phenomena are frequently referred to in the literature as adsorption or specific adsorption.

7.1.4 Ionization and Adsorption of Humic Substances

Biogenic substances within the soil or water or sediment environments will transform to humic substances, which are important organic substances in natural waters. In sea water, humic substance constitutes 6~30% of its total organic substances. Swamp water appears yellow because it contains a great quantity of humic substances. The organic substances contained in river water which is not polluted are mainly humic substances.

Humic substances can be grouped into three types according to their solubility. The constituent which is soluble not only in acid but also in base is termed fulvic acid; the constituent which is soluble in base but not in acid is termed humic acid; the constituent which is insoluble not only in acid but also in base is termed humin.

Humic substances are high-molecular-weight compounds of a complex nature. Their molecular weights are in the range of $300 \sim 1 \times 10^6$. Among them, the molecular weight of fulvic acid is about in the range of hundreds to thousands; the molecular weight of humic acid is about in the range of thousands to several tens of thousands. Generally speaking, fulvic acid has a molecular

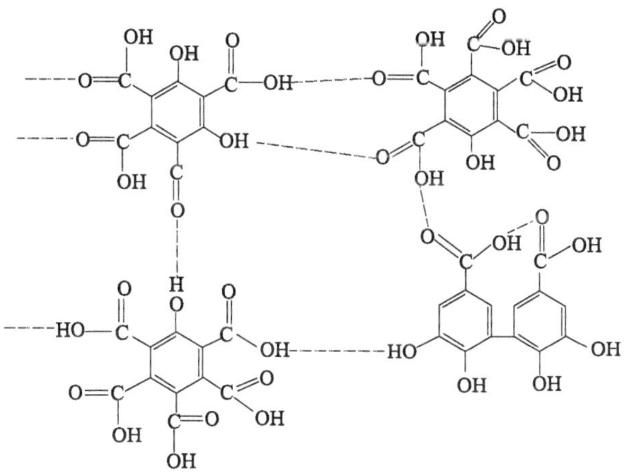

FIGURE 7.4 **Structure of fulvic acid.**

weight lower than humic acid and humin; the deeper the color, the greater the molecular weight.

Humic substances mainly consist of carbon, hydrogen, oxygen, nitrogen, and a little phosphorus. Structurally, humic substances may be considered to be largely condensation products of phenols, quinones, and amino compounds. They consist of polyanions containing aromatic nuclei with functional groups (OH and COOH) linked through a kind of atomic bridge (—O—, —CH—, —CH_2—, ═CH—, —S—S—, and hydrogen bond). Fulvic acid contains more hydrophilic functional groups. A typical building block may contain structures such as those in Fig. 7.4.

Humic substances have porous and loose structure and great specific surface area, just like a sponge. Since humic substances have various functional groups, they are amphoteric polyelectrolytes. The ionization and adsorption of their functional groups will lead the different charges at their surfaces. The sign of charge depends on the pH value of the solution. When the pH value is lower, positive charge is dominant; when pH value is higher, negative charge is dominant. There must be a pH value at which the net charge equals zero, which is the isoelectric point (pzc), just like hydrous oxide minerals.

7.1.5 Amphoteric Behavior of Protein

The water polluted by domestic sewage commonly contains proteins. Proteins consist of various amino acids, which are amphoteric molecules. Proteins acquire their charge mainly through the ionization of carboxyl and adsorption of amino groups to give COO^- and NH_3^+, respectively. The ionization and adsorption of these groups, and so the net molecular charge, depends strongly on the pH value of the solution. At low pH value a protein molecule will be positively charged, and at high pH value it will be negatively charged. The pH at which the

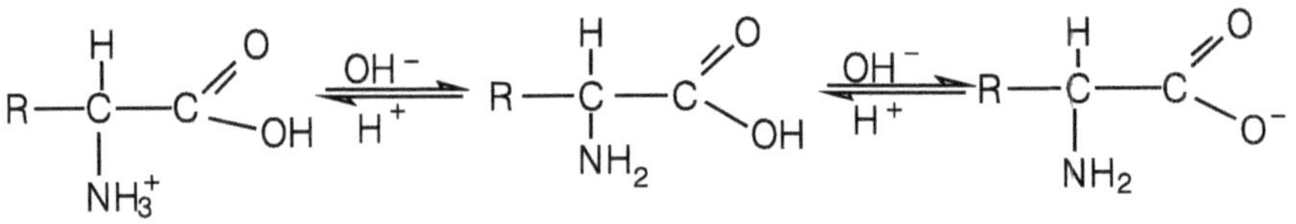

FIGURE 7.5 Changes of protein charges with pH value.

net charge is zero is the pzc value. The change of protein charge with pH value is shown in Fig. 7.5. There are different pzc values for different amino acids; for example, the pzc value of pepsase is 1.1, casein is 3.7, RNase is 9.5, and lysozyme is 11.0, which are all in the range of 2~11.

An introduction to some mechanisms for colloids to acquire electrical charges has been presented previously. Although not all cases are included, these mechanisms are very common in water dispersion systems. From the view of thermodynamics, there is only one reason, that is, the high dispersion degree of colloidal systems causes the rise of surface-free energy of particles, thus the particles have to reduce their surface-free energy through ionization and adsorption; as a result, they obtain the charges.

7.2 ELECTROKINETIC PHENOMENA

When studying the electrical properties of colloid systems, scientists discovered electrokinetic phenomena. Based on electrokinetic phenomena, scientists then realized the structure of the electric double layers of colloids. Therefore electrokinetic phenomena are of great importance. The four electrokinetic phenomena are introduced in the following sections.

7.2.1 Electrophoresis

Electrophoresis is the movement of a charged surface plus attached material relative to stationary liquid by an applied electric field. Fig. 7.6 illustrates a simple electrophoresis cell, which is a U-type tube with an electrode at each end.

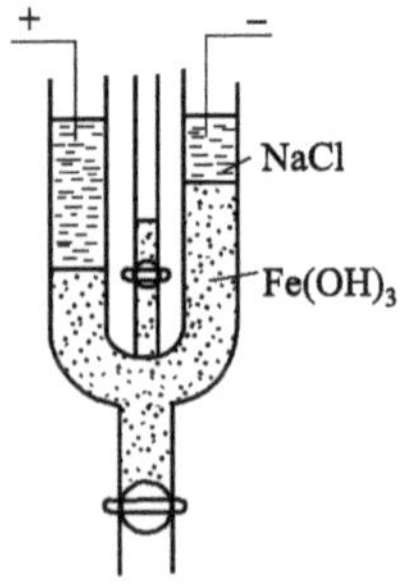

FIGURE 7.6 Apparatus for electrophoresis experiment.

TABLE 7.1 Comparison of Electrophoresis Velocities Between Colloidal Particles and Ions

Ions (particles)	Electrophoresis velocities (10^{-6} m s^{-1}/100 v m^{-1})
H^+	32.6
OH^-	18.0
Na^+	4.5
K^+	6.7
Cl^-	6.8
$C_3H_7COO^-$	3.1
$C_8H_{17}COO^-$	2.0
Colloid	2.0~4.0

In the electrophoresis experiment with this apparatus, first let us add a brownish red sol of $Fe(OH)_3$ to the apparatus, then a colorless solution of NaCl onto the tops of the sol of each end, respectively. This must be added carefully so that the clear interface between the sol and the solution form at both ends. After the electricity is connected for some time, we observe that the brownish red sol of $Fe(OH)_3$ falls down at the anode end, but rises up at the cathode end. This phenomenon proves that the brownish red sol of $Fe(OH)_3$ is positively charged. By the same experiment, we know that the sol of As_2S_3 is negatively charged.

The electrophoresis velocity of colloidal particles is proportional to the charge quantity of the particle and the electric potential gradient of the electric field, but inversely proportional to the viscosity of the medium and the size of the particle. Although the sizes of colloidal particles are much greater than ions, it was proven that the electrophoresis velocities of colloidal particles are approximately the same as the ions in magnitude, as shown in Table 7.1, which indicates that the charge quantity of colloidal particles is quite great.

7.2.2 Electroosmosis

Electroosmosis is the movement of liquid relative to a stationary charged surface by an applied electric field. Fig. 7.7 illustrates a simple apparatus for studying electroosmotic flow through a porous plug. This apparatus is also a U-type tube with an electrode at each end. In an electroosmosis experiment, first let us add an electrolyte solution to the apparatus, then connect electricity for some time; it will be observed that the surface of the solution rises up at the cathode end, but falls down at the anode end.

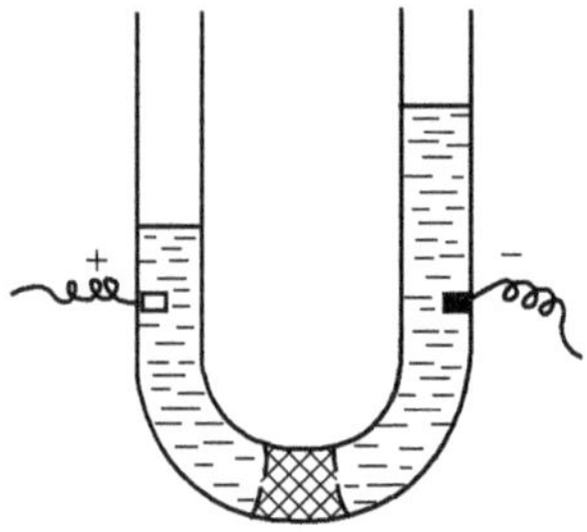

FIGURE 7.7 **Apparatus for electroosmosis experiment.**

7.2.3 Streaming Potential

Streaming potential is opposite to electroosmosis. When an electrolyte solution is forced to flow through a stationary capillary or a porous plug, an electrical potential difference will be produced between the two ends of capillary or porous plug. This electrical potential difference is referred to as streaming potential, which means the liquid carries away the ions with the charge opposite to the surface, giving rise to a streaming current, consequently, leading to the accumulations of charge at the ends and an electrical field. Fig. 7.8 illustrates an apparatus for studying streaming potential.

When streaming potential is measured, a very high resistance measuring instrument, such as a vibrating capacitor electrometer, must be used to minimize current drain. Most of the difficulties associated with streaming potential measurement originate at the electrodes. A superimposed asymmetry potential often develops; however, by reversing the direction of liquid flow this asymmetry potential can be made to reinforce and oppose the streaming potential and can, therefore, be allowed.

7.2.4 Sedimentation Potential

Sedimentation potential is due to the electric field created by charged particles settling in a gravity field or a centrifugal force field. This means when settling, the charged particles leave the ions with the charge opposite them in their

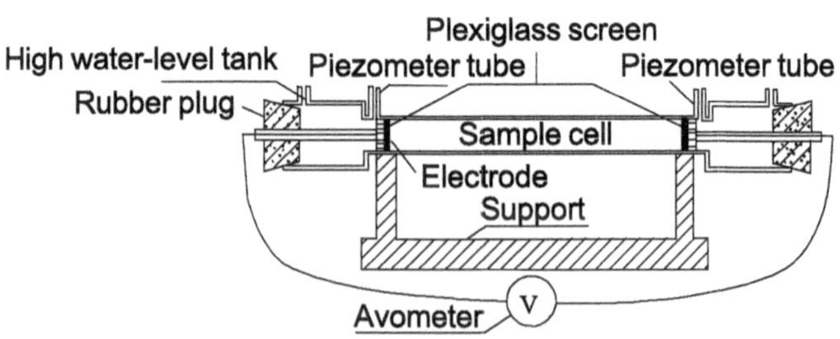

FIGURE 7.8 **Apparatus for measuring streaming potential.**

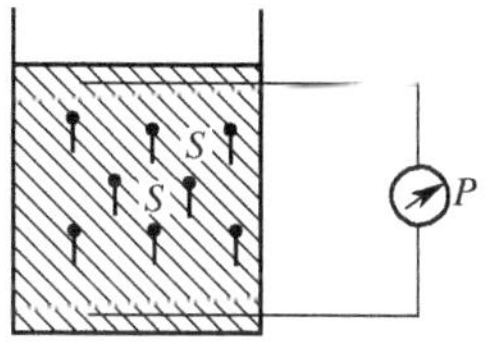

FIGURE 7.9 Apparatus for studying sedimentation potential.

original spot, thus making the positive charges and negative charges separate from each other. The apparatus for studying sedimentation potential is shown in Fig. 7.9, where *P* represents potentiometer and *S* represents colloidal particles.

In summary, from among the four electrokinetic phenomena introduced previously, electrophoresis and electroosmosis are relative movements between solid phase and liquid phase, which are caused by electrical potential difference applied to the system, thus termed "the movement caused by electricity," while streaming potential and sedimentation potential are the electrical potential differences caused by the relative movements between solid phase and liquid phase, thus termed "the electricity caused by movement." Electrokinetic phenomena imply that the dispersed phase and dispersion medium have the opposite electrical charges, predicting the electric double layer existing between them.

7.3 MODEL OF ELECTRIC DOUBLE LAYER

The surface charge of dispersed phase influences the distribution of nearby ions in the dispersion medium. Ions of opposite charge (counterions) are attracted toward the surface, and ions of like charge (coions) are repelled away from the surface. This, together with the mixing tendency of thermal motion, leads to the formation of an electric double layer made up of the charged surface and a neutralizing excess of counterions over coions. The theory of the electric double layer deals with this distribution of ions and hence with the magnitude of the electric potentials, which occur in the locality of the charged surface. This is a necessary first step toward understanding many of the experimental observations concerning the electrokinetic properties, stability, etc., of charged colloidal systems. Many kinds of structures of electric double layer have been developed up until now, but the representative models are the Helmholtz, Gouy–Chapman, and Stern.

7.3.1 Helmholtz Model of Parallel-Plate Capacitor

Early in 1879, Helmholtz proposed the parallel-plate capacitor model of the electric double layer to explain electrokinetic phenomena. According to his theory, the electric double layer resembles a capacitor, as shown in Fig. 7.10, in which the surface of particles is shown as a plane. In this model, the surface

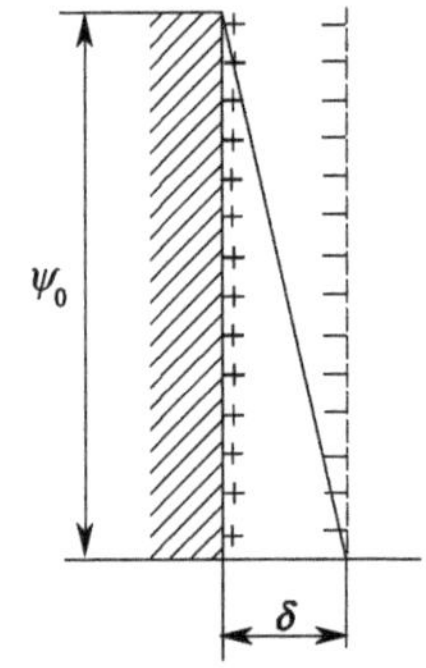

FIGURE 7.10 **Helmholtz model of parallel-plate capacitor.**

charges constitute one plate of the capacitor, while the counterions parallelly arranged constitute the other plate of the capacitor. The distance between these two plates (layers) is very small, approximately equalling the radius of the ion. The electric potential linearly decreases from one plate to the other plate, just as with the parallel-plate capacitor, the relation between the surface electric potential ψ_0 and the surface electric density σ_0 is given by

$$\sigma_0 = \frac{\varepsilon\psi_0}{\delta} \tag{7.7}$$

where δ is termed as the distance between two layers, and ε is termed as the dielectric constant.

The explanation of the Helmholtz model of parallel-plate capacitors for electrokinetic phenomena is as follows: under the effect of applied electric field or driving force, the charged particle and the counterions are separated from each other and move toward opposite electrodes or directions respectively, thus leading to electrokinetic phenomena. This model played some important rule for the early research of electrokinetic phenomena, but it cannot distinguish surface potential (thermodynamic electric potential) ψ_0 from electrokinetic or ζ (*zeta*) potential. The latter research shows that the bounded water layer moving with the particle is much thicker than that of the Helmholtz electric double layer, that is, the counterions are included inside the bounded water layer; therefore, it will not be possible for the electrokinetic phenomena to take place because the electric double layer is electroneutral as an entirety.

7.3.2 Gouy–Chapman Model of Diffuse Double Layer

Aimed at the problems of the Helmholtz model discussed previously, Gouy and Chapman indicated that the distribution of counterions are influenced by both electrical force and random thermal motion at the same time; therefore, the counterions are diffusely distributed, not parallelly arranged, forming a diffuse double layer at equilibrium state, as shown in Fig. 7.11.

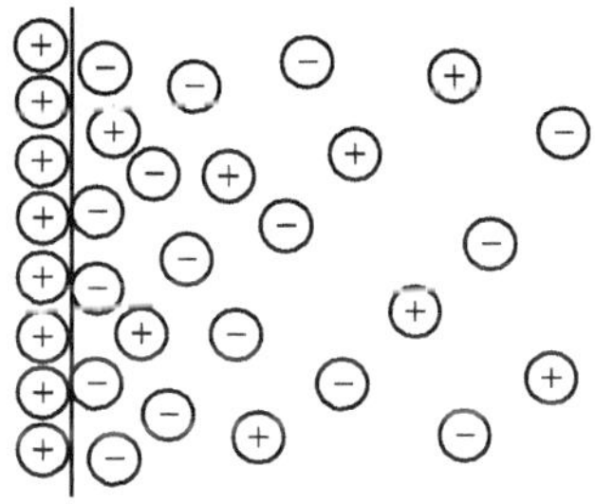

FIGURE 7.11 **Goüy–Chapman model of diffuse double layer.**

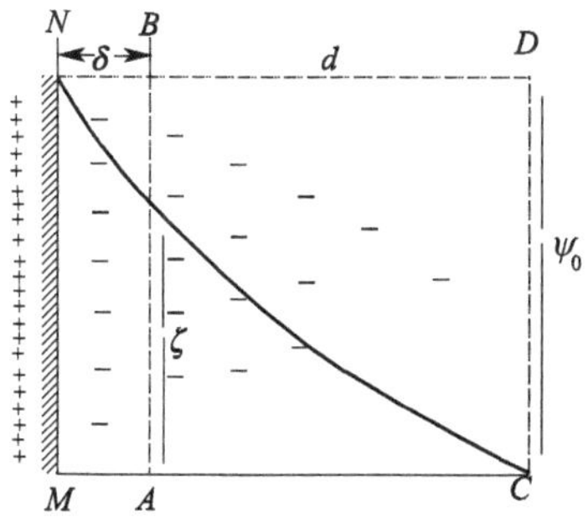

FIGURE 7.12 **Change of the electric potential of diffuse double layer.**

It can be seen from Fig. 7.11 that the concentration of excess counterions is higher nearby the particle surface and decreases with the increase in the distance from surface to the solution until equaling to that of coions. Fig. 7.12 illustrates that the electric potential of diffuse double layer also decreases with the increase in the distance from the surface to the solution.

In this figure ψ_0 is the electric potential at the surface and ψ is at a distance x from the surface. It is well known that a bound water layer (the layer of liquid immediately adjacent to a particle) moves with the particle in electrokinetic phenomena. This means the boundary of this bound water layer is just the interface between particle and liquid in their relative movement, which is known as the surface of shear. We define the potential at the surface of shear to be the zeta potential ζ, which is also termed electrokinetic potential because it can be determined by the electrokinetic experiment. In Fig. 7.12 AB represents the surface of shear, and zeta potential ζ is indicated.

The simplest quantitative treatment of the diffuse double layer is that due to Gouy and Chapman, which is based on the following assumptions.

1. The surface is assumed to be flat, of infinite extent, and uniformly charged.
2. The ions in the double layer are assumed to be point charges distributed according to the Boltzmann distribution.

3. The solvent is assumed to influence the double layer only through its dielectric constant, which is assumed to have the same value throughout the double layer.
4. A single symmetrical electrolyte of charge number z will be assumed. This assumption facilitates the derivation whilst losing little owing to the relative unimportance of the coion charge number.

Taking the surface to be positively charged (Figs. 7.11 and 7.12) and applying the Boltzmann distribution

$$n_+ = n_{0+}\exp(-ze\psi/K_BT) \tag{7.8}$$

$$n_- = n_{0-}\exp(ze\psi/K_BT) \tag{7.9}$$

where n_+ and n_- are the respective numbers of positive and negative ions per unit volume at the points where the potential is ψ (ie, where the electric potential energy is $ze\psi$ and $-ze\psi$, respectively), and n_{0+} and n_{0-} are the corresponding bulk concentration of each ionic species, e is the electric quantity of one electron, and K_B is the Boltzmann constant. Eqs. (7.8) and (7.9) show that the concentration of counterions is different from the concentration of coions, and the former is greater than the latter, which is shown in Fig. 7.13.

The net volume charge density ρ at the points where the potential is ψ, therefore, is given by

$$\begin{aligned}\rho &= ze(n_+ - n_-)\\ &= zen_0[\exp(-ze\psi/K_BT) - \exp(ze\psi/K_BT)]\\ &= -2n_0ze\sin h(ze\psi/K_BT)\end{aligned} \tag{7.10}$$

Also ρ is related to ψ by Poisson's equation as follows:

$$\nabla^2\psi = -\frac{\rho}{\varepsilon} \tag{7.11}$$

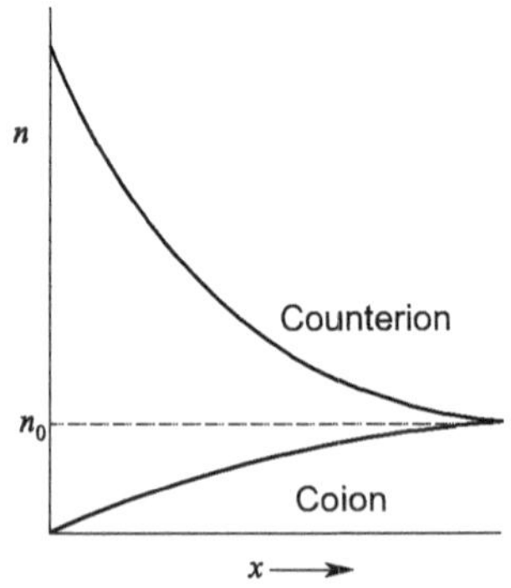

FIGURE 7.13 Distribution of ions in diffusion electric double layer.

where ∇^2 is the Laplace operator, which equals to $\left(\frac{\partial^2}{\partial x^2}+\frac{\partial^2}{\partial y^2}+\frac{\partial^2}{\partial z^2}\right)$. For a flat double layer

$$\nabla^2\psi = \frac{d^2\psi}{dx^2}$$

thus

$$\frac{d^2\psi}{dx^2} = -\frac{\rho}{\varepsilon} \tag{7.12}$$

The combination of Eqs. (7.10) and (7.12) gives

$$\frac{d^2\psi}{dx^2} = \frac{2n_0}{\varepsilon} ze \cdot \sin h(ze\psi / K_B T) \tag{7.13}$$

Eq. (7.13) is not easy to solve, but if the surface potential is very low so that $ze\psi_0/K_BT \ll 1$, the solution will be greatly simplified. For example, when temperature is 25°C and $\psi_0 \ll 25.7$ mv, $\sinh(ze\psi/K_BT) \approx ze\psi/K_BT$, thus

$$\frac{d^2\psi}{dx^2} = \frac{2n_0}{\varepsilon K_B T} z^2 e^2 \psi = \kappa^2 \psi \tag{7.14}$$

where

$$\kappa = \left(\frac{2e^2 n_0 z^2}{\varepsilon K_B T}\right)^{1/2} = \left(\frac{2e^2 N_A c_z^2}{\varepsilon K_B T}\right)^{1/2} \tag{7.15}$$

where N_A is the Avogadro constant and c is the concentration of electrolytes. Solving Eq. (7.14) gives

$$\psi = \psi_0 e^{-\kappa X} \tag{7.16}$$

For spherical particles, by similar mathematical operation we obtained

$$\psi = \psi_0 \frac{a}{r} e^{-\kappa(r-\alpha)} \tag{7.17}$$

where α is the radius of particles, and r is the distance from the center of the sphere.

Eqs. (7.16) and (7.17) are very important results, which show that at low potential the potential decreases exponentially with the distance from the charged surface, and the velocity of the decrease depends on the size of K. It is known that K is a very important physical quantity, and its reciprocal is of dimension of the length.

By equating the surface charge with the net space charge in the diffuse part of the double layer, we have

$$\sigma_0 = -\int_0^\infty \rho\, dx \tag{7.18}$$

At very low potential, applying Eqs. (7.12) and (7.16) leads to

$$\begin{aligned}\sigma_0 &= \varepsilon \int_0^\infty \frac{d^2\psi}{dx^2} dx = \varepsilon \left(\frac{d\psi}{dx}\right)_{x=\infty} - \varepsilon \left(\frac{d\psi}{dx}\right)_{x=0} = -\varepsilon \left(\frac{d\psi}{dx}\right)_{x=0} \\ &= \varepsilon \kappa \psi_0 = \frac{\varepsilon \psi_0}{\kappa^{-1}}\end{aligned} \tag{7.19}$$

Comparing Eq. (7.19) with Eq. (7.7), we can see that, at low potentials, a diffuse double layer has the same capacity as a parallel plate condenser with a distance $1/\kappa$ between the plates. It is customary to refer to $1/\kappa$ (the distance over which the potential decreases by an exponential factor at low potentials) as the "thickness" of the diffuse layer. If the $1/\kappa$ is smaller, that is, the diffuse double layer is thinner, the potential will decrease by an exponential factor more quickly, which indicates that the velocity of the potential decrease depends on the value of κ.

Eq. (7.15) shows κ is proportional to $n_0^{1/2}$ and z. For an aqueous solution of a symmetrical electrolyte at 25°C, Eq. (7.15) becomes

$$\kappa = 0.328 \times 10^{10} \left(\frac{cz^2}{\text{mol dm}^{-3}}\right)^{1/2} \text{m}^{-1} \tag{7.20}$$

For a 1–1 electrolyte the double layer thickness is, therefore, about 1 nm for 10^{-1} mol dm^{-3} solution and about 10 nm for 10^{-3} mol dm^{-3} solution. For unsymmetrical electrolytes the double layer thickness can be calculated by taking z to be the counterion charge number. If remembering the aforementioned values, we will able to estimate the values of κ at different c and z conveniently.

Eqs. (7.15) and (7.20) show that the increasing concentration or charge number of electrolytes will lead to the increase in the value of κ, that is, the decrease in the thickness of the double layer (the double layer is compressed); as a result, the electrical potential will drop down with the distance from the surface more quickly, thus leading to the decease in ζ potential. This also can be understood from Eq. (7.16), which shows the velocity of the decrease of the potential depends on the size of K. The effects of concentration and charge number of electrolytes are illustrated in Fig. 7.14a, b, respectively.

As we have discussed previously, Eq. (7.16) has been deduced for surfaces with very low surface potential. For surface potential that is not very low, we must solve the Eq. (7.13) without approximation. Here let us omit mathematical operations, and only give the final expressions as follows.

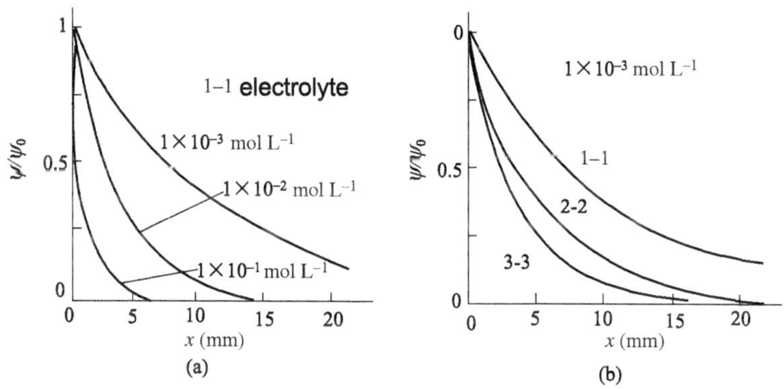

FIGURE 7.14 Effects of concentration and charge number on potential drop. (a) Effect of concentration; and (b) effect of charge number.

$$\gamma = \gamma_0 e^{-\kappa x} \tag{7.21}$$

where

$$\gamma = \frac{\exp(ze\psi/2K_BT) - 1}{\exp(ze\psi/2K_BT) + 1} \quad \gamma_0 = \frac{\exp(ze\psi_0/2K_BT) - 1}{\exp(ze\psi_0/2K_BT) + 1} \tag{7.22}$$

It is not easy to observe the relationship between ψ_0 and ψ from Eq. (7.21), but it can be simplified in some special situations:

1. If ψ_0 is very low so that $ze\psi_0/2K_BT \ll 1$ ($K_BT/e = 25.6$ mv at 25°C), the Debye–Hückel approximation $\exp(ze\psi_0/2K_BT) \approx 1 + ze\psi_0/2K_BT$ can be made, thus $\gamma_0 \approx ze\psi_0/4K_BT$ and $\gamma \approx ze\psi/4K_BT$; therefore, Eq. (7.21) will become Eq. (7.16).
2. When ψ_0 is not very low, ψ must be very low at the spot far from surface ($\kappa x > 1$), thus the γ in Eq. (7.21) can be approximately substituted by $ze\psi/4K_BT$; therefore, we have

$$\psi = \frac{4K_BT}{ze}\gamma_0 e^{-\kappa x} \tag{7.23}$$

Eq. (7.23) indicates that no matter how high the surface potential is, the potential always decreases exponentially with the distance from the charged surface.

1. When ψ_0 is very high so that $ze\psi_0/2K_BT \gg 1$, $\gamma_0 \approx 1$, thus Eq. (7.23) can be further simplified to

$$\psi = \frac{4K_BT}{ze} e^{-\kappa x} \tag{7.24}$$

Eq. (7.24) indicates that the potential at the spot far from the surface is independent of ψ_0.

Gouy–Chapman theory overcame the defects of Helmholtz model and distinguished surface potential (thermodynamic electric potential) ψ_0 from electrokinetic or ζ (*zeta*) potential, thus making the explanation for electrokinetic phenomena more reasonable; in addition, it obtained the expressions for charge distribution and potential distribution in electrical double layer from the Poisson–Boltzmann relation. Despite all this, there are still some experiment facts which conflict with Gouy–Chapman theory, for instance,

1. If the concentration of electrolyte in solution is not very low (eg, 0.1 mol dm^{-3} 1–1 electrolyte), and the potential nearby surface is quite high (eg, 200 mv), then the concentration of counterions calculated from Eqs. (7.8) and (7.9) will reach up to 240 mol dm^{-3} in this spot, which is obviously not possible.
2. Although the Gouy–Chapman theory distinguished surface potential ψ_0 from ζ potential, it did not give ζ potential the definite physical meaning. According to the Gouy–Chapman model, although ζ potential decreases with the increase in the concentration of ions, it always keeps the same sign as that of the surface, and its limiting value is zero. But in some cases, it is discovered that ζ potential increases with the increase in the concentration of ions, and sometimes becomes opposite in sign. This fact cannot be explained by the Gouy–Chapman theory.

7.3.3 Stern Model

The treatment of the diffuse double layer outlined in the previous section is based on an assumption of point charges in the electrolyte medium. In 1924, Stern proposed

1. Ions have a definite size, and the center of an ion can only approach the surface to within its hydrated radius;
2. Ions are attached (albeit temporarily) to the surface by electrostatic and/or van der Waals forces strong enough to overcome thermal agitation. In recent years, the study proved that there are some "specific adsorptions" between ions and surfaces, which are nonelectrostatic forces including covalent bond, coordination bond, and hydrogen bond.

According to the Stern model, the double layer is divided into two parts separated by a plane (the Stern plane) located at about a hydrated ion radius from the surface. The centers of any adsorbed ions are located in the Stern layer, that is, between the surface and the Stern plane. Ions with centers located beyond the Stern plane form the diffuse part of the double layer, for which the Gouy–Chapman treatment outlined in the previous section, with ψ_0 replaced by ψ_δ (the potential at the stern plate), is considered to be applicable.

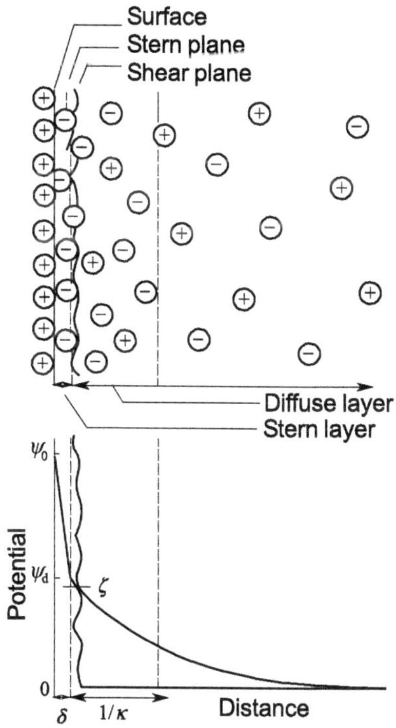

FIGURE 7.15 **Stern electric double layer.**

The potential linearly changes with the distance from ψ_0 to ψ_δ in the Stern layer, just as in the Helmholtz model, and decays exponentially with the distance from ψ_δ to zero in the diffuse double layer just as in the Gouy–Chapman model. Actually, the Stern model was obtained by combining the Helmholtz model and the Gouy–Chapman model, as shown in Fig. 7.15.

The variable ψ_δ can be estimated from the electrokinetic measurement. Electrokinetic behavior depends on the potential at the surface of shear between the charged surface and the electrolyte solution. This potential is referred to as the electrokinetic or ζ (zeta) potential, as discussed in Section 7.3.2. The exact location of the shear plane is another unknown feature of the electric double layer. In addition to ions in the Stern layer, a certain amount of solvent will probably be bound to the charged surface and form a part of the electrokinetic unit. It is, therefore, reasonable to suppose that the shear plane is usually located at a small distance further out from the surface than the Stern plane and that ζ is, in general, marginally smaller in magnitude than ψ_δ (Fig. 7.15). It is customary to assume the identity of ψ_δ and ζ, and the bulk of experimental evidence suggests that errors introduced through this assumption are generally small, especially at lyophobic surfaces.

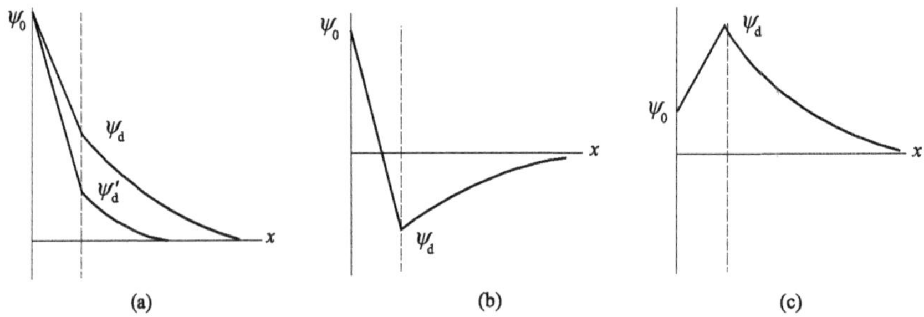

FIGURE 7.16 Effects of electrolytes on electric double layer. (a) Stern layer are in equilibrium with the ions in the diffuse layer; (b) specific adsorption takes place; (c) adsorption of surface active coions.

The Stern model overcame the drawback of previous models and explained problems which cannot be explained by previous models, making the electrical double layer theory more scientific and more complete, as follows.

1. Since the ions in the Stern layer are in equilibrium with the ions in the diffuse layer, the increase of concentration or charge number of electrolyte in solution will make more counterions enter the Stern layer, leading to a quicker potential drop in the Stern layer, thus resulting in the decrease in ζ potential and compression of thickness of double layers, as shown in Fig. 7.16a.
2. When specific adsorption takes place, it is possible, especially with polyvalent or surface active counterions, for reversal of charge to take place within the Stern layer, that is, for ψ_0 and ψ_δ to have opposite signs, as shown in Fig. 7.16b.
3. Adsorption of surface active coions could create a situation in which ψ_δ has the same sign as ψ_0 and is greater in magnitude, as shown in Fig. 7.16c.
4. Since the potential drops from ψ_0 to ψ_δ in the Stern layer, the potential in the diffuse layer will not be very high, thus the counterions concentration calculated from Eqs. (7.8) and (7.9) will not reach up to an unreasonable value.

7.4 ELECTROKINETIC THEORY AND EXPERIMENT

Current colloid and interface chemistry gives us a reasonable explanation for electrokinetic phenomena: dissociation and adsorption of substances at the interface between solid and liquid cause the charge excess at the solid surface, making the distribution of ions in solution nearby the solid surface become non-uniform, thus forming an electric double layer. When the electric double layer is agitated by an external force, the relative movement between adsorption layer and diffuse layer takes place, thus producing a series of electrokinetic phenomena. Based on the structure of electric double layer established in earlier section, the scientific explanation and quantitative treatment for electrokinetic phenomena can be made respectively as follows.

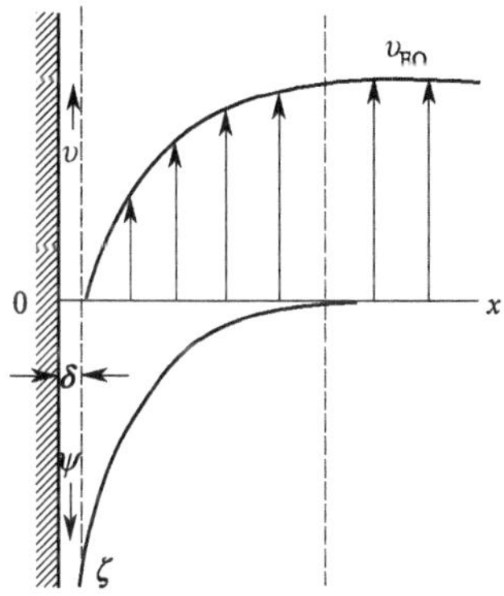

FIGURE 7.17 **Distribution of velocity in electroosmosis flow.**

7.4.1 Theory and Experiment of Electroosmosis

7.4.1.1 Theory of Electroosmosis

Porous plugs can be seen as a bunch of capillaries; let us consider the electroosmosis flow which takes place in one of the capillaries.

Fig. 7.17 shows that an electric field is applied tangentially along the charged surface of a capillary tube, and there are some net electric charges due to excess of counterions in diffuse layers. These ions carrying liquid along with them move in appropriate directions under the electric field force, thus causing the electroosmosis flow. The velocity starts to gradually increase from $x = \delta$ (outer edge of shear plane) and reaches the maximum at $\psi = 0$, then remains unchanged. Since the net electric charge is zero outside the double layer, there is no effect of electric field and no velocity gradients in the bulk solution. Suppose there is one volume unit with thickness dx and area A at the distance x from the surface in the electric double layer. The forces applied to this volume unit are: electric field force, viscosity force of external liquid, and viscosity force of internal liquid, as shown in Fig. 7.18.

Let the charge density be ρ, electric field strength be E, and flow velocity be v. When the electroosmosis flow becomes stable, the electric field force and viscosity force are equal, therefore,

$$E\rho A dx + \eta A\left(\frac{dv}{dx}\right)_{x+dx} - \eta A\left(\frac{dv}{dx}\right)_{x} = 0 \tag{7.25}$$

$$E\rho dx + \eta\left[\left(\frac{dv}{dx}\right)_{x+dx} - \left(\frac{dv}{dx}\right)_{x}\right] = 0$$

$$E\rho dx + \eta\left(\frac{d^2v}{dx^2}\right)dx = 0 \tag{7.26}$$

$$E\rho = -\eta\left(\frac{d^2v}{dx^2}\right)$$

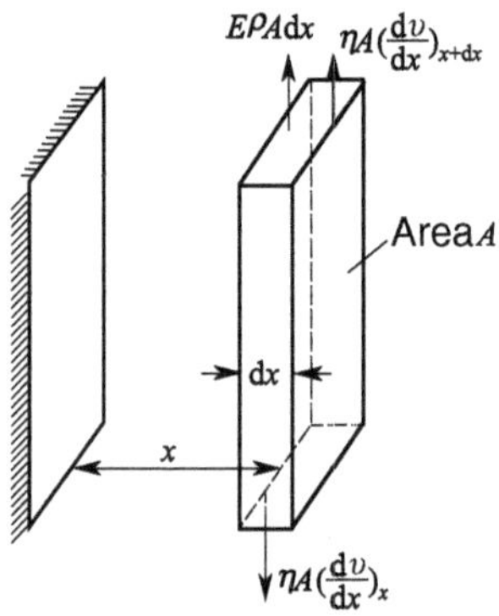

FIGURE 7.18 **Analysis of the forces applied to the volume unit.**

For a plat particle

$$\frac{d^2\psi}{dx^2} = -\frac{\rho}{\varepsilon} \tag{7.12}$$

Substituting Eq. (7.12) into Eq. (7.26) gives

$$\varepsilon E \frac{d^2\psi}{dx^2} = \frac{d}{dx}\left(\eta \frac{dv}{dx}\right)$$

$$\varepsilon E \frac{d^2\psi}{dx^2} dx = d\left(\eta \frac{dv}{dx}\right)$$

Integrating the above equation leads to

$$\varepsilon E \frac{d\psi}{dx} = \eta \frac{dv}{dx} + c$$

where c is the integrating constant. Now let us consider the region outside the double layer. In this region $d\psi/dx = dv/dx = 0$, therefore, C = 0; thus we have

$$\varepsilon E \frac{d\psi}{dx} = \eta \frac{dv}{dx}$$

$$dv = \frac{\varepsilon E}{\eta} d\psi$$

$$\int_{\delta}^{x} dv = \frac{\varepsilon E}{\eta} \int_{\varsigma}^{\psi} d\psi$$

$$v = \frac{\varepsilon E}{\eta}(\psi - \zeta) \tag{7.27}$$

Since $\psi = 0$ and v remains unchanged in the region outside the double layer, the flow velocity in the region outside the double layer is given by

$$v_\infty = -\frac{\varepsilon}{\eta} E\zeta \tag{7.28}$$

Since the thickness of the double layer is very small, Eq. (7.28) approximately represents the flow velocity in the tube provided the sign is omitted. At the unit electric field strength

$$v_{EO} = \frac{v_\infty}{E} = \frac{\varepsilon\zeta}{\eta} \tag{7.29}$$

from which ζ potential can be calculated.

7.4.1.2 Experimental Measurements of Electroosmosis

7.4.1.2.1 Volume Method

Commonly, the volume flow velocity is experimentally measured. Since the radius of the capillary is much greater than the thickness of the double layer, we do not consider the difference of flow velocity between the double layer and the main body of the tube, and regard Eq. (7.28) as the expression of electroosmosis flow velocity in a whole capillary tube. For a porous plug in electric fields, the volume of outflow liquid per unit time is given by

$$Q = v_\infty A = \frac{\varepsilon E\zeta}{\eta} A \tag{7.30}$$

where A is the sum of cross-sectional areas of all capillaries in the porous plug. In order to obtain the value of A, let conductance be G, resistance R, specific conductance λ, and the length of cell L, then we have

$$G = \frac{1}{R} = \lambda \frac{A}{L} \tag{7.31}$$

thus

$$R = \frac{L}{A} \cdot \frac{1}{\lambda} \tag{7.32}$$

Let the voltage be V and the current I, then we have

$$V = IR \tag{7.33}$$

Substituting Eq. (7.32) into Eq. (7.33) gives

$$V = I \cdot \frac{L}{A} \frac{1}{\lambda}$$

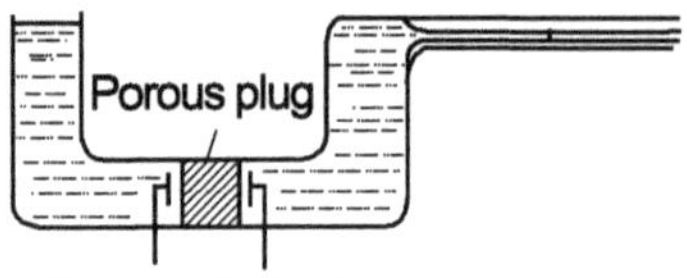

FIGURE 7.19 **Apparatus for the experiment measurement of electroosmosis.**

or

$$A \cdot \frac{V}{L} = \frac{I}{\lambda}$$

or

$$A \cdot E = \frac{I}{\lambda} \tag{7.34}$$

where E is the electric field strength. Substituting Eq. (7.34) into Eq. (7.30) yields

$$Q = \frac{\varepsilon \zeta I}{\eta \lambda} \tag{7.35}$$

or

$$\zeta = \frac{\eta \lambda}{\varepsilon I} Q \tag{7.36}$$

It can be seen from Eq. (7.36) that the ζ potential of the wall of capillaries in the porous plug can be calculated based on the data from the experiment's measurements.

The apparatus for the experiment's measurements of electroosmosis is shown in Fig. 7.19. It can be seen from this apparatus that the volume of outflow liquid can be obtained by reading the scales on the level thin tube on the right.

7.4.1.2.2 Reverse Pressure Method

If the thin tube from which liquid flows out in Fig. 7.19 is not level, but made to be vertical; the electroosmosis flow will cause different altitudes of liquid in the two columns of apparatus, thus leading to a pressure difference between two columns. This pressure difference will in turn cause a reverse pressure flow which is in the opposite direction of the electroosmosis flow. With the electroosmosis proceeding, the pressure difference becomes greater and greater so that the electroosmosis flow will be offset by the reverse pressure flow, as shown in Fig. 7.20, and the system will reach equilibrium at last. The pressure difference between two columns at equilibrium is termed balance reverse pressure.

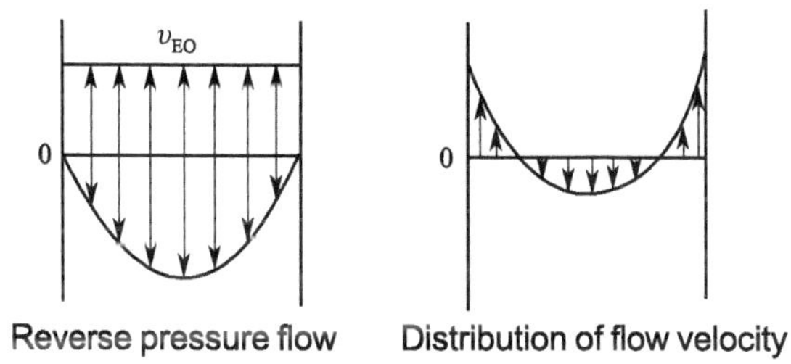

FIGURE 7.20 **Distribution of flow velocity in reverse pressure method.**

Let the balance reverse pressure be P; the reverse pressure flow can be obtained by the Poiseuille equation (Eq. 6.7) as follows.

$$\overleftarrow{Q} = \frac{\pi r^4 P}{8\eta L} \tag{7.37}$$

where r and L are the radius and length of the capillary tube, respectively. According to Eq. (7.30), the electroosmosis flow is given by

$$\vec{Q} = \frac{\varepsilon E\zeta}{\eta} \cdot \pi r^2 \tag{7.38}$$

At equilibrium $\vec{Q} = \overleftarrow{Q}$, leading to

$$P = \frac{8\varepsilon E\zeta L}{r^2} = \frac{8\varepsilon\zeta V}{r^2} \tag{7.39}$$

It can be seen from Eq. (7.39) that the balance reverse pressure only depends on the voltage applied to the apparatus, but does not depend on the length of the capillary tube. Since r is the average radius of capillary tubes in the porous plug, it does not depend on the size of the plug either. Therefore after measuring the balance reverse pressure P, the ζ potential of the wall of the porous plug can be calculated as follows.

$$\zeta = \frac{Pr^2}{8\varepsilon V} \tag{7.40}$$

7.4.2 Theory and Experiment of Electrophoresis

7.4.2.1 Theory of Electrophoresis

7.4.2.1.1 Smoluchowski Equation

In the research of the electric double layer, the shape of the double layer is commonly described in terms of a dimensionless parameter $\kappa\alpha$ which can be considered to be the ratio of particle radius α to the thickness of double layer κ^{-1}. The Smoluchowski theory deals with the case of $\kappa\alpha \gg 1$, which implies that the particle is greater and the double layer is thinner; therefore, the surface of a particle can be treated as a plane, then Eqs. (7.28) and (7.29) can be directly used for electrophoresis because both electroosmosis and electrophoresis are the relative motion between solid phase and liquid phase. Therefore, the electrophoresis velocity is given by

$$v_E = \frac{\varepsilon E \zeta}{\eta} \tag{7.41}$$

It is customary to refer to the velocity of particles at unit electric field strength as the electrophoretic mobility.

$$u_E = \frac{\varepsilon \zeta}{\eta} \tag{7.42}$$

Eq. (7.42) is referred to as the Smoluchowski equation, which shows the relation between electrophoresis velocity and ζ potential for $\kappa\alpha \gg 1$. Although the SI unit of electrophoretic mobility is m s^{-1}/v m^{-1} (or m^2/v s), μm s^{-1} (or cm.s^{-1}) is commonly used as the unit for v_E, and v cm^{-1} for E, thus the unit of electrophoretic mobility becomes μm s^{-1}/v cm^{-1} (cm^2 s^{-1}/v).

7.4.2.1.2 Hückel Equation

Consider $\kappa\alpha$ to be small enough ($\kappa\alpha \ll 1$) for a spherical particle to be treated as a point charge in an unperturbed electric field, but let the particle be large enough for Stokes' law to apply. Equating the electrical force on the particle with the frictional resistance of the medium,

$$QE = 6\pi\eta\alpha v_E \tag{7.43}$$

where Q is the electrical quantity on the particle, E the electric field strength, η the viscosity of medium, α the radius of the particle, and v_E the electrophoretic velocity.

The zeta potential is the resultant potential at the surface of shear due to the charge $+Q$ of the electrokinetic unit and $-Q$ of the mobile part of the double layer, that is,

$$\begin{aligned}\zeta &= \frac{Q}{4\pi\varepsilon\alpha} - \frac{Q}{4\pi\varepsilon\left(\alpha+\kappa^{-1}\right)} \\ &= \frac{Q}{4\pi\varepsilon\alpha(1+\kappa\alpha)}\end{aligned} \tag{7.44}$$

where ε is the dielectric constant of the electrolyte medium. Since $\kappa\alpha \ll 1$, neglecting $\kappa\alpha$ gives

$$Q = 4\pi\varepsilon\alpha\zeta \tag{7.45}$$

Substituting Eq. (7.45) into Eq. (7.43) gives the electrophoretic mobility

$$u_E = \frac{v_E}{E} = \frac{\varepsilon\zeta}{1.5\eta} \tag{7.46}$$

Eq. (7.46) is referred to as the Hückel equation. This equation is not likely to be applicable to particle electrophoresis in aqueous media; for example, particles of radius 10^{-8} m suspended in 1–1 aqueous electrolyte solution would require an electrolyte concentration as low as 10^{-5} mol. dm^{-3} to give $\kappa\alpha = 0.1$. The equation, however, does have possible applicability to electrophoresis in nonaqueous media of low conductance.

7.4.2.1.3 Henry Equation

Comparing Eq. (7.46) with Eq. (7.42), we know the relation between electrophoretic mobility and ζ potential depends on the value of $\kappa\alpha$. Henry indicated that the electric field of double layer and the applied electric field are superimposed, thus causing mutual distortion, which will affect the electric mobility of particles. After careful analysis and detailed calculation of the effect of $\kappa\alpha$ on the interaction between the electric field of double layer and the applied electric field, Henry derived a general electrophoretic equation for conducting and nonconducting spheres which takes the form

$$u_E = \frac{\varepsilon\zeta}{1.5\eta}[1 + KF(\kappa\alpha)] \tag{7.47}$$

where $F(\kappa\alpha)$ varies between zero for small values of $\kappa\alpha$ and 1.0 for large values of $\kappa\alpha$, and $K = (\lambda_0 - \lambda_p)/(2\lambda_0 + \lambda_p)$, where λ_0 is the specific conductance of the bulk electrolyte solution and λ_p is the specific conductance of the particle. Eq. (7.47) is referred to as the Henry equation.

For common nonconducting particle ($K = 1/2$) the Henry equation can be written in the form

$$u_E = \frac{\varepsilon\zeta}{1.5\eta}\left[1 + \frac{1}{2}F(\kappa\alpha)\right] \tag{7.48}$$

or

$$u_E = \frac{\varepsilon\zeta}{1.5\eta}f(\kappa\alpha) \tag{7.49}$$

The correction factor $f(\kappa\alpha)$ changes with the values of $\kappa\alpha$, as shown in Fig. 7.21.

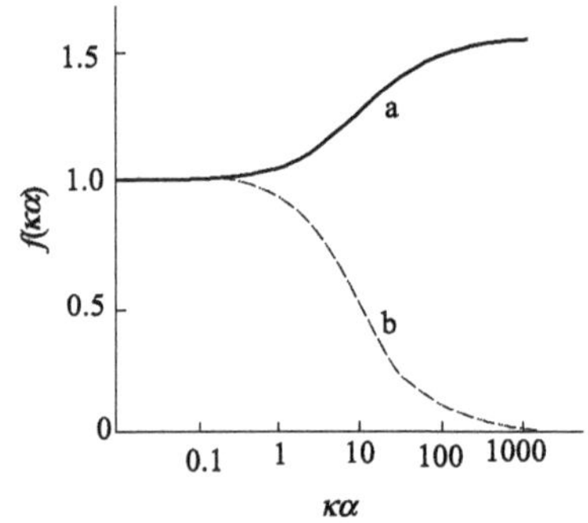

FIGURE 7.21 Change of correction factor with the values of $\kappa\alpha$. a, Nonconducting spherical particle; and b, conducting spherical particle.

It can be seen from Fig. 7.21 that for the case of $\kappa\alpha \ll 1$, we have $F(\kappa\alpha) = 0$ and $f(\kappa\alpha) = 1$, thus the Henry equation becomes the aforementioned Hückel equation; for the case of $\kappa\alpha \gg 1$, we have $F(\kappa\alpha) = 1$ and $f(\kappa\alpha) = 1.5$, thus the Henry equation becomes the aforementioned Smoluchowski equation. Therefore, the Hückel equation and Smoluchowski equation represent two limiting situations of the Henry equation, respectively.

For common conducting particles, $\lambda_p \gg \lambda_0$, thus K approaches -1. When $\kappa\alpha$ is very small, $F(\kappa\alpha)$ will approach zero; therefore, the Hückel equation will still be applicable to the calculation of electrophoresis mobility. When $\kappa\alpha$ is very large, $f(\kappa\alpha)$ of the conducting particle changes with $\kappa\alpha$ and will approach zero, as shown by the dashed line in Fig. 7.21. Metal sole seems to be such a case, but its surface will quickly become polarized in an electric field, therefore, it can still be treated as a nonconducting particle.

The counterions in the diffuse layer of a double layer show a net movement in a direction opposite to that of the particle under the influence of the applied electric field. This creates a local movement of liquid which opposes the motion of the particle and is known as electrophoretic retardation. It is allowed for in the Henry equation.

The movement of the particle relative to the diffuse part of the double layer results in the double layer being distorted, thus the center of the positive charge and the center of the negative charge do not coincide because a finite time (relaxation time) is required for the original symmetry to be restored by diffusion and conduction. The resulting asymmetric diffuse part of the double layer forms an additional electric field in the opposite direction, which exerts an additional retarding force on the particle, known as the relaxation effect, and this is not accounted for in the Henry equation. This retarding effect depends on the ζ potential, the value of $\kappa\alpha$, the charge number, the mobility of ions, etc. It can be neglected when $\kappa\alpha < 0.1$ or $\kappa\alpha > 300$, but must be accounted for when $\kappa\alpha$ is in the range of $0.1 \sim 300$.

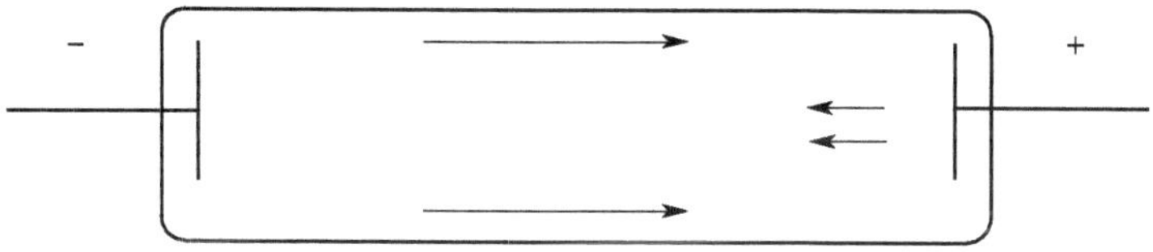

FIGURE 7.22 Schematic diagram of an electrophoresis cell.

7.4.2.2 Experiment Measurements of Electrophoresis

7.4.2.2.1 Microscope Electrophoresis

If the material under investigation is in a state of reasonably stable suspension containing microscopically visible particles or droplets, then the electrophoretic behavior can be observed and measured directly. The electrophoresis cell usually consists of a horizontal glass tube, of either rectangular or circular cross-section, with an electrode at each end. Fig. 7.22 is the schematic diagram of an electrophoresis cell. If the concentration of salt is lower than 1×10^{-2} mol dm^{-3}, a platinum black electrode will be more convenient; if the concentration of salt is higher, it will be better to use reversible electrodes (eg, Cu/$CuSO_4$ or Ag/AgCl) to avoid polarization.

Electrophoretic measurements by the microscope method are influenced by the simultaneous occurrence of electroosmosis. The internal glass surface of the cell is generally charged, causing an electroosmosis flow of liquid near to the tube walls, together with a compensating return flow of liquid with maximum velocity at the center of the tube because the cell is closed. The distribution of velocity is the same as that in Fig. 7.20b. The true electrophoretic velocity is observed only at the locations in the tube where the electroosmosis flow and return flow of the liquid cancel each other. For a cylindrical cell the "stationary level" is located at 0.146 of the internal diameter from the cell wall. For a flat cell the "stationary levels" are located at fractions of about 0.2 and 0.8 of the total depth, the exact location depending on the width/depth ratio.

The operations of the microscopic electrophoresis measurement are very simple and quick; in addition, the sample amount needed is very small, and the measurement can be made under the original environment of particles themselves. Therefore, this method is commonly used to measure the ζ potential of the current dispersion system. But only the microscopically visible particles or droplets can be measured by microscopic electrophoresis. If the particle is too small or is a charged macromolecule, the method of moving boundary electrophoresis should be used instead of the microscopic electrophoresis method.

7.4.2.2.2 Moving Boundary Electrophoresis

An alternative electrophoretic technique is to study the movement of a boundary formed between a sol or solution and pure dispersion medium. Fig. 7.6 is the

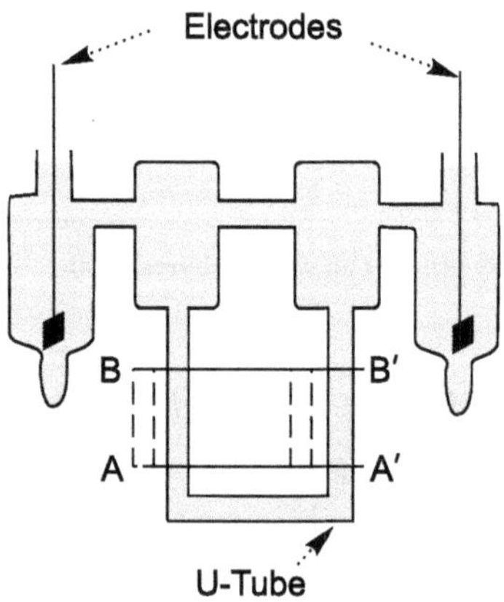

FIGURE 7.23 **Arne Tiselius electrophoresis cell.**

schematic diagram of a simple apparatus for moving boundary electrophoresis. When using this apparatus, it is very important and difficult to keep the interface clear and undisturbed during the operation. The apparatus most commonly used is that of Arne Tiselius, which is illustrated schematically in Fig. 7.23. The Tiselius moving boundary method has found wide application, not only for measuring ζ potential, but also, particularly, for separating, identifying, and estimating dissolved macromolecules, particularly proteins. The apparatus consists of a U-tube of rectangular cross-section, that is, segmented in such a way that the sections between the lines AA′ and BB′ in Fig. 7.23 can be laterally displaced with respect to the rest of the apparatus. The offset segments of the U tube are filled with the colloidal dispersion, and after thermal equilibration with the buffer solution contained in other parts of the apparatus; the various sections are aligned so that sharp boundaries are obtained. The location of the boundaries is usually observed by schlieren optics, which identify refractive index gradients (see Section 3.2).

As the macroions migrate in the electric field, the schlieren peak becomes displaced and the mobility of a colloidal component may be determined by measuring the rate of boundary movement per unit electric field, thus its ζ potential may be calculated, for example, when $\kappa\alpha \gg 1$; from Eq. (7.41), we have

$$\zeta = \frac{\eta v}{\varepsilon E} \tag{7.50}$$

Customarily, we use V. cm^{-1} as the unit of E, in which V is the SI unit, but cm is the cgs unit. We also use poise (cgs unit) as the unit of η, and cm. s^{-1} (cgs unit) as the unit of v. Therefore, we must do the unit conversion in calculation as follows

$$\zeta = \frac{\eta v}{\varepsilon E \frac{1}{300}} \quad \text{(statvolt)}$$
$$= \frac{\eta v}{\varepsilon E \frac{1}{300}} \times 300 \text{ (V)} \tag{7.51}$$
$$= \frac{\eta}{\varepsilon} u_E \times 300^2 \text{ (V)}$$

Substituting the aforementioned customary units into Eq. (7.51) will give the ζ potential of the SI units (V or mV).

Example 7.1 Electrophoresis and Electroosmosis

Calculate the electroosmosis flow velocity of the KCl solution through a quartz membrane. Suppose we knew that the electrokinetic potential of the quartz particle $\zeta_0 = 30 \times 10^{-3}$ V, which is calculated from the electrophoresis velocity of quartz particles in the same solution without considering the electrophoretic retardation, and $\eta = 1 \times 10^{-3}$ Pa s, $\varepsilon_r = 81$, $E = 2 \times 10^2$ V m^{-1}, $I = 2 \times 10^{-2}$ A, $\lambda = 2 \times 10^{-2}\ \Omega^{-1}\cdot\text{m}^{-1}$, $\alpha = 1 \times 10^{-7}$ m, and $\kappa = 5 \times 10^7$ m^{-1}.

Solution

From the data given previously we have

$$\kappa\alpha = 5.0$$

Without considering the mutual distortion of the electric double layer and the applied electric field due to their superposition, we use the Smoluchowski equation

$$u_E = \frac{\varepsilon.\zeta_0}{\eta}$$

thus

$$\zeta_0 = \frac{u_E \eta}{\varepsilon_r \varepsilon_0}$$

When considering the mutual distortion of the electric double layer and the applied electric field due to their superposition, we use the Henry equation. According to $\kappa\alpha = 5.0$, we obtained $f(\kappa\alpha) = 1.16$ by consulting the relevant table in the reference.

Thus

$$u_E = \frac{\varepsilon\zeta}{1.5\eta} f(\kappa a)$$

$$\zeta = \frac{1.5\eta u_E}{\varepsilon_r \varepsilon_0 f(\boldsymbol{\kappa} a)} = \frac{1.5\zeta_0}{f(\boldsymbol{\kappa} a)} = \frac{1.5 \times 30 \times 10^{-3}}{2 \times 1.16} = 38 \times 10^{-3} \text{ (V)}$$

Substituting it into Eq. (7.35) gives

$$Q=\frac{\varepsilon\zeta I}{\eta\lambda}=\frac{38\times10^{-3}\times2\times10^{-2}\times81\times8.85\times10^{-12}}{1\times10^{-3}\times2\times10^{-2}}=2.72\times10^{-8}\ (\mathrm{m^3 s^{-1}})$$
$$=2.72\times10^{-2}(\mathrm{cm^3 s^{-1}})=1.63\ (\mathrm{cm^3 min^{-1}})$$

7.4.3 Theory and Experiment of Streaming Potential

When a liquid is forced by a pressure to flow through a stationary capillary or porous plug, the liquid will carry away the counterions with the charge opposite to the surface, giving rise to a streaming current I_s, thus forming an electric field due to the charge accumulation at the two ends of capillary or porous plug, which will in turn produce a current I_c opposite to I_s in direction. When I_s equals I_c, the electric potential difference between the two ends is referred to as streaming potential. From Poiseuille equation (Eq. 6.7) and diffusion theory, the following equation was derived:

$$\frac{E_s}{p}=\frac{\varepsilon\zeta}{\eta}\cdot\frac{1}{\lambda} \tag{7.52}$$

where E_s is streaming potential, λ is specific conductance. Rearranging Eq. (7.36) gives

$$\frac{Q}{I}=\frac{\varepsilon\zeta}{\eta}\cdot\frac{1}{\lambda} \tag{7.53}$$

Comparing Eq. (7.52) with Eq. (7.53), we know that both streaming potential and electric-osmosis flow velocity (they are opposite phenomena) are expressed by equations of the same form. The electric-osmosis flow velocity produced by unit current is equivalent to the potential difference produced by unit pressure. Both of them are independent of the size of capillary or porous plug. It can be seen from Eq. (7.52) that the value of streaming potential is inversely proportional to the specific conductance of the medium.

Since the specific conductance of hydrocarbon is several orders of magnitude smaller than that of water, the streaming potential produced is very high when this kind of liquid is pumped; as a result, it is easy to cause spark under high pressure, leading to fire disaster. Therefore some necessary measures should be taken in order to avoid the accident; for example, adding anelectrolyte which is soluble in oil or making the equipment well grounded.

Example 7.2 Streaming Potential

The relative dielectric constant of fuel in the effuser of the jet plane is 8.0, and the viscosity is 3×10^{-3} Pa s (0.03 poise). When pumped with the pressure of 3 atm, the ζ potential between the wall of the effuser and the oil is 125 mV. Suppose the oil has ions of very low concentration and is equivalent to NaCl solution of 10^{-8} mol. dm^{-3}. Make the necessary assumptions, and calculate the streaming potential produced between the two ends of the effuser.

Solution

Let the relative dielectric constant be ε_r, the dielectric constant of the vacuum be ε_0, the molar conductance be Λ_m, and the limiting molar conductance be Λ_m^∞, respectively. From the aforementioned data we have

$$\varepsilon = \varepsilon_r \cdot \varepsilon_0 = 8 \times 8.854 \times 10^{-12}$$

$$\zeta = 0.125(\text{V})$$

$$p = 101{,}325 \times 3\ (\text{Pa})$$

$$\eta = 3 \times 10^{-3}(\text{Pa s})$$

$$\begin{aligned} \lambda &\approx c \times \Lambda_m(\text{NaCl}) \approx c \times \Lambda_m^\infty(\text{NaCl}) \\ &= c \times \left[\lambda_m^\infty(\text{Na}^+) + \lambda_m^\infty(\text{Cl}^-)\right] \\ &= 10^{-8} \times 10^3 (50.11 + 76.34) \times 10^{-4} (\text{Sm}^{-1}) \\ &= 1.2645 \times 10^{-7} (\text{Sm}^{-1}) \end{aligned}$$

Substituting all of the aforementioned items into Eq. (7.52) gives

$$E_s = \frac{\varepsilon \zeta p}{\eta \lambda} = \frac{8 \times 8.854 \times 10^{-12} \times 0.125 \times 101{,}325 \times 3}{3 \times 10^{-3} \times 1.2645 \times 10^{-7}} = 7{,}094.8\ (\text{V})$$

7.4.4 Theory and Experiment of Sedimentation Potential

We can use the equation of streaming potential to discuss sedimentation potential. If $\kappa\alpha \gg 1$, Eq. (7.52) can be directly used to calculate sedimentation potential, but the pressure P should be the gravity force applied to the particles as follows:

$$p = \frac{4}{3}\pi a^3 (\rho_1 - \rho_0) n_0 g \tag{7.54}$$

where α is the radius of particle, ρ_1 and ρ_0 the density of particles and liquid medium respectively, and n_0 the particle number per unit volume. Substituting Eq. (7.54) into Eq. (7.52) gives the sedimentation potential E_{sd} as follows:

$$E_{sd} = \frac{4\pi a^3 (\rho - \rho_0) n_0 g \varepsilon \zeta}{3\eta\lambda} \tag{7.55}$$

As mentioned previously, this equation is suitable for the condition $\kappa\alpha \gg 1$. In general, it should be multiplied by correction factor *f*, which is similar to the treatment in the case of electrophoresis.

$$E_{sd} = \frac{4\pi a^3(\rho - \rho_0)n_0 g\varepsilon\zeta}{3\eta\lambda} f \tag{7.56}$$

where *f* is a function of $\kappa\alpha$, and the quantitative relation is the same as that of electrophoresis.

Example 7.3 Sedimentation Potential

Calculate the sedimentation potential of barium carbonate particles in the aqueous solution of NaCl. Suppose the total volume of particles $V = 0.2\ m^3, \varepsilon_r = 81$, $\zeta = 40 \times 10^{-3}$ V, $\rho - \rho_0 = 2.1 \times 10^3$ kg m^{-3}, $\eta = 1 \times 10^{-3}$ Pa s, $\lambda = 1 \times 10^{-2}$ S m^{-1}.

Solution

Substituting all of the aforementioned data gives

$$\begin{aligned} E_{sd} &= \frac{4\pi a^3(\rho - \rho_0)n_0 g\varepsilon\zeta}{3\eta\lambda} \\ &= \frac{V(\rho - \rho_0)g\varepsilon_r\varepsilon_0\zeta}{\eta\lambda} \\ &= \frac{0.2 \times 2.1 \times 10^3 \times 9.8 \times 81 \times 8.85 \times 10^{-12} \times 40 \times 10^{-3}}{1 \times 10^{-3} \times 1 \times 10^{-2}} \\ &= 11.8 \times 10^{-3} (V) \end{aligned}$$

7.5 COAGULATION THERMODYNAMICS: DLVO THEORY OF COLLOID STABILITY

Coagulation is an important colloidal phenomenon that changes the size distribution of particles from a large number of small particles to a small number of large particles. This phenomenon concerns the interactions between particles. There is van der Waals attractive force between two colloidal particles, which make the particles approach each other. When the particles are close enough, the repulsive force will arise from the overlapping of the diffuse parts of the double layers around the particles. In any specific system it is the relative magnitude of the attractive and repulsive forces between the particles that governs the stability of the system. Because both of these forces are relative to the interparticle distance, they can be expressed in terms of potential energy. Derjaguin and Landau and Verwey and Overbeek have independently developed a quantitative theory in which the stability of lyophobic sols is treated in terms of the potential energy changes, which take place when particles approach each other. This theory is the famous DLVO theory.

7.5.1 van der Waals Attractive Energy Between Particles

Colloidal particles can be considered as an assembly of a large number of molecules. Three types of such intermolecular attraction are recognized:

1. Two molecules with permanent dipoles mutually orientate one another in such a way that, on average, attraction results.
2. Dipolar molecules induce dipoles in other molecules so that attraction results.
3. Attractive forces are also operative between nonpolar molecules. These universal attractive forces (known as dispersion forces) are due to the polarization of one molecule by fluctuations in the charge distribution in a second molecule and vice versa.

With the exception of highly polar materials, the dispersion forces account for nearly all of the van der Waals attraction, which is operative. The attractive energy between two molecules is very short range, varying inversely with the sixth power of the intermolecular distance. Since all of the aforementioned interactions are attraction, they are expressed in terms of negative value. Hamaker assumed that the interaction between two particles equals the sum of intermolecular interactions. The results of such summations predict that the attractive energy between colloidal particles decays much less rapidly than that between individual molecules. For two parallel plates, the interaction energy per unit area is given by

$$V_A = -\frac{A}{12\pi D^2} \tag{7.57}$$

where D is the distance between the two plates, and A is the Hamaker constant, which depends on the interaction parameters of molecules that constitute the particle. For the case of two identical spheres of radius α with the shortest distance H between the spheres not greater than c. 10–20 nm, and when $H \ll a$, the attraction energy V_A is given by the approximate expression as follows:

$$V_\mathrm{A} = -\frac{Aa}{12H} \tag{7.58}$$

Eqs. (7.57) and (7.58) are applicable if the radii of particles are much greater than the distances between particles. Most actual colloid systems meet this condition. If the radii of particles are very small, the correction for the plate thickness δ and sphere radius a must be considered; thus the corresponding equations become

$$V_\mathrm{A} = -\frac{A}{12\pi}\left[\frac{1}{D^2} + \frac{1}{(D+2\delta)^2} - \frac{2}{(D+\delta)^2}\right] \tag{7.59}$$

and

$$V_A = -\frac{A}{12}\left[\frac{4a^2}{H^2+4aH}+\frac{4a^2}{(H+2a)^2}+2\ln\frac{H^2+4aH}{(H+2a^2)}\right] \tag{7.60}$$

The Hamaker constant A is a very important parameter, which directly influences the size of V_A. There are two methods of calculating A, one is the microcosmic method, which starts from molecule properties, for example, polarizability and ionization energy. The other is the macrocosmic method, in which particles and medium are considered to be a continuous phase, and A is obtained from the varying of dielectric property with frequency. Table 7.2 presents Hamaker constants for some materials of practical interest.

Since the results measured by different methods are different, the values of Hamaker constants in the table lie in some ranges. This shows the accurate calculation and experimental measure of the Hamaker constant are still tasks to be solved in the future. The Hamaker constant has the unit of energy, and the values for common materials are in the range of $10^{-19}\sim10^{-20}$ J. The presence of liquid dispersion medium, rather than a vacuum (or air), between the particles notably lowers the attraction energy. The constant A in Eqs. (7.57) and (7.58) must be replaced by an effective Hamaker constant, calculated from the expression

$$A_{131} = (A_{11}^{1/2} - A_{33}^{1/2})^2 \tag{7.61}$$

where subscripts 11 and 33 refer to particles and dispersion medium, respectively. A_{131} is the effective Hamaker constant of the particle in dispersion medium. Eq. (7.61) indicates that van der Waals interaction will always be the attraction of each other and be weakest when the particles and the dispersion medium are chemically similar, since A_{11} and A_{33} will be of similar magnitude and the value of A_{131} will, therefore, below.

TABLE 7.2 Hamaker Constants for Selected Materials

Material	$A/10^{-20}$ J (macrocosmic method)	$A/10^{-20}$ J (microcosmic method)
Water	3.0~6.1	3.3~6.4
Ion crystal	5.8~11.8	15.8~41.8
Metals	22.1	7.6~15.9
Quartz	8.0~8.8	11.0~18.6
Hydrocarbon	6.3	4.6~10
Polystyrene	5.6~6.4	6.2~16.8

Example 7.4 van der Waals Attractive Energy Between Particles

Calculate the attractive potential energy between two identical spherical particles. Suppose the particles have radius of 1×10^{-7} m, and coasted with 2.0-nm thick lipoid monolayer. There is 2.0-nm thick medium of water between the particles coated. In addition, thermodynamic temperature $T = 300$ K, and Hamaker constant $A = 6 \times 10^{-20}$ J. The effect of lipoid monolayer on attraction can be ignored.

Solution

From Eq. (7.58) for two identical spheres

$$V_A = -\frac{Aa}{12H} = -\frac{6\times10^{-20}\times1\times10^{-7}}{12\times(2+2+2)\times10^{-9}} = -8.3\times10^{-20}(\text{J})$$

Since $1K_BT = 1.38\times10^{-23}\times300 = 0.414\times10^{-20}(\text{J})$,

$$V_A = -20K_BT.$$

7.5.2 Double Layer Repulsive Energy Between Particles

If the diffuse layer is not separated from the charged surface, the particle as an entirety will be electrically neutral; therefore, there will be no repulsive force between two particles. When two particles approach close to the overlapping of the diffuse part of the double layer, repulsive force will take place. The simplest method for calculating the repulsive energy of electric double layer is Langmuir, which is introduced as follows. Fig. 7.24 illustrates the overlap of two double layers when a pair of identical plates with surface potential ψ_0 is brought to a surface separation D.

The profile of the potential drop before overlap is shown by the dashed line, and the net potential variation after overlap is shown as the solid line. Since the surface potential of the two plates are identical, the curve of net potential $\psi(x)$ must be symmetrical in the area between the two plates after

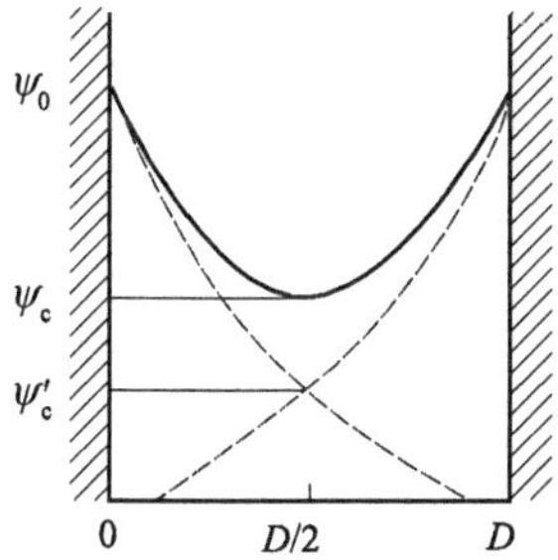

FIGURE 7.24 Schematic representation of the overlap of two double layers.

overlap and reach the minimum ψ_c at $D/2$ spot (let the potential at this spot before overlap be ψ_c'). The concentrations of ions after overlap should be different from that before overlap. According to the Boltzmann distribution, at $D/2$ spot they are

$$n_+ = n_{0+} \exp(-ze\psi_c/K_BT) \quad (7.62)$$

and

$$n_- = n_{0-} \exp(ze\psi_c/K_BT) \quad (7.63)$$

where n_+ and n_- are the number concentrations of the cations and the anions at $D/2$ spot respectively, n_0 the number concentration of the cations or the anions outside the region between the two plates, and e the electrical quantity of an electron. Thus the sum of concentration of cation and anion at $D/2$ spot is

$$n = 2n_0 \cos h(ze\psi c/K_BT) \quad (7.64)$$

But in the solution outside the region between two plates, the sum of concentration of cation and anion $n = 2n_0$ (and $\psi = 0$). The ion concentration inside the region is greater than that outside the region. This different ion concentration will cause osmosis pressure, leading to the repulsive force between the two plates. The osmosis pressure (see chapter: Osmotic Pressure) is given by

$$\pi = cRT \quad (7.65)$$

where π is osmosis pressure and c is mole concentration. Substituting the concentration difference for concentration in the Eq. (7.65) gives the repulsive force per area of plate as follows:

$$p = 2cRT[\cos h(ze\psi_c/K_BT) - 1] \quad (7.66)$$

Again substituting the number concentration n_0 for the mol. concentration and Boltzmann constant for the gas constant in Eq. (7.66) and integrating it from $D/2$ to ∞ will give the repulsive potential energy as follows:

$$V_R = +2\int_{x=D/2}^{\infty} p\,dx = -2\int_{\infty}^{x=D/2} 2n_0K_BT[\cos h(ze\psi_c/K_BT) - 1]dx \quad (7.67)$$

which is corresponding to the work to be done by transferring the particles from $x = D/2$ to the infinity point. Since $\cos h(ze\psi_c/K_BT) > 1$, the value of V_R is always positive. But the relation between $\cos h(ze\psi_c/K_BT)$ and the distance x from the surface is very complicated; therefore, Eq. (7.67) is difficult to solve. If the extent of overlap of double layers is not very high, $\kappa(D/2) > 1$, and both of ψ_c and ψ'_c are small, we will have $\psi_c \approx 2\psi'_c$ and $\cos h(ze\psi_c/K_BT) \approx 1 + \frac{1}{2}(ze\psi_c/K_BT)^2$.

In addition, we know from Eq. (7.24) that $\psi = (4K_BT/ze)\,\gamma_0\exp(-\kappa x)$ at the spot far from the surface no matter how high the ψ_0 is, thus

$$\psi_c' = \frac{4K_BT}{ze}\gamma_0 e^{-\kappa x} \tag{7.68}$$

$$\psi_c = \frac{8K_BT}{ze}\gamma_0 e^{-\kappa x} \tag{7.69}$$

Substituting these items discussed previously into Eq. (7.67) and integrating it gives

$$V_R = \frac{64n_0K_BT}{\kappa}\gamma_0^2 e^{-2\kappa x} \tag{7.70}$$

or

$$V_R = \frac{64n_0K_BT}{\kappa}\gamma_0^2 e^{-\kappa D} \tag{7.71}$$

where V_R is the repulsive potential energy at unit cross section caused by overlap of double layers. From Eq. (7.22), we know

$$\gamma_0 = \frac{\exp(ze\psi_0/2K_BT - 1}{\exp(ze\psi_0/2K_BT) + 1}$$

It shows that the repulsive potential energy is related to ψ_0 only through γ_0. When ψ_0 is very high, γ_0 approaches 1. In this case, V_R is nearly independent of ψ_0 and only depends on the concentration and charge number of electrolytes.

As far as spherical particles are concerned, the situation will be very complicated. Currently, we are only able to solve the equation in some special cases, for example, when $\kappa x \gg 1$ and the extent of overlap of double layers is very low, the repulsive potential energy between two spherical particles is given by

$$V_R = \frac{64n_0K_BT}{\kappa^2}\pi a\gamma_0^2 e^{-\kappa H} \tag{7.72}$$

where α is the radius of the particle and H the shortest distance between two particles.

7.5.3 Total Energy of Interaction Between Particles

The total energy of interaction between particles is obtained by the summation of the attraction and repulsion energies, that is, $V_T = V_A + V_R$, which is illustrated in Fig. 7.25. The general character of the resultant potential energy–distance

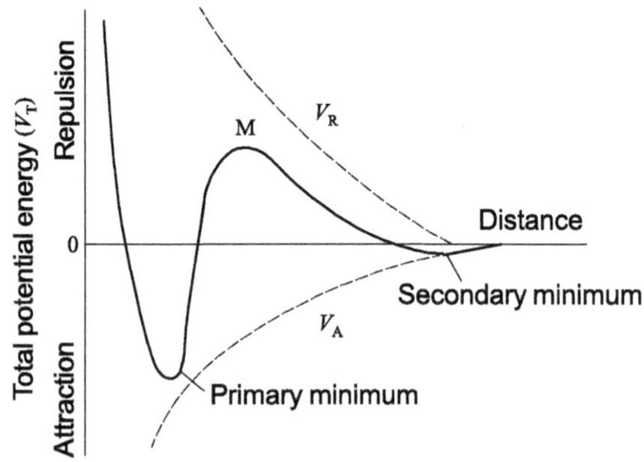

FIGURE 7.25 **Total interaction energy curves.**

curve can be deduced from the properties of the two forces. On the one hand, the repulsion energy is an exponential function of interparticle distance, as shown in Eq. (7.72), which quickly rises up with the decrease in interparticle distance and reaches a limiting value at last; that is, it drops down with the increase in the interparticle distance and reaches zero quickly. On the other hand, the attraction energy as an inverse power of the interparticle distance, as shown in Eq. (7.58), slowly rises up with the decrease in the interparticle distance and reaches infinite at last; that is, drops down with the increase in the interparticle distance and reaches zero slowly. Consequently, van der Waals attraction predominates at small and large interparticle distances respectively so that V_T becomes negative. At intermediate distances double layer repulsion predominates so that V_T becomes positive, depending on the actual values of two forces. When the interparticle distance is extremely small, the attraction energy drops down sharply due to overlapping of electron clouds (Born repulsion), thus forming a deep minimum termed the primary minimum. When the distance is large enough, the repulsion energy quickly drops down to zero, but the attraction energy slowly drops down to zero, thus forming the secondary minimum. Between the primary minimum and the secondary minimum the V_R exceeds the V_A, thus forming a potential energy maximum, which is termed the potential barrier. If the potential barrier is higher compared with the thermal energy K_BT of the particles, the system should be stable; otherwise, the system will be unstable and coagulate at last.

Eq. (7.58) shows that V_A only depends on the Hamaker constant A. Since A is unchangeable for a certain system, the attraction potential energy is also unchangeable, thus its curve keeps constant during the process. But V_R changes with ζ potential and the distance at which repulsion force displays (ie, double layer thickness $1/\kappa$), thus leading to the corresponding change of total potential energy curve. Fig. 7.26 illustrates the changes of total potential energy curves with the change in the values of κ. When increasing the concentration

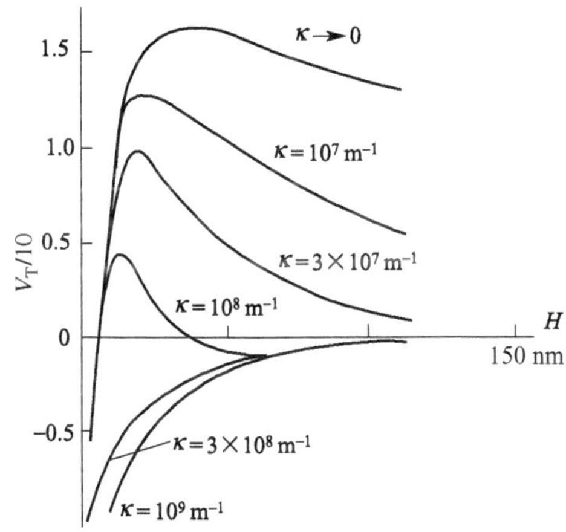

FIGURE 7.26 Changes of total potential energy curve with κ.

of electrolytes or the charge number of counterions, κ increases, leading to the quicker decrease of V_R with the increase in the interparticle distance, and making V_A become relatively stronger and V_R display only at the shorter distance; therefore, the height of the energy barrier falls down and moves toward the left until all curve of total potential energy is below the axis.

When two particles approach each other, they first reach the secondary minimum. Because the secondary minimum is smaller compared with the Brownian motion energy of the particles, the very weak flocculation takes place at this distance and the flocs formed are very easy to break due to disturbance. When the particles continue to approach each other, they will reach the energy barrier. If the height of the energy barrier is lower compared with the Brownian motion energy of the particles, the particles will go over the energy barrier, approaching further, then the van der Waals attraction will become predominate, and the particles will strongly attract each other at the primary minimum because the primary minimum is much deeper compared with the Brownian motion energy of the particles; as a result, the coagulation will take place. Generally speaking, the energy barrier of stable sole reaches up to thousands of K_BT, but the average kinetic energy of particles is only $(3/2)K_BT$; therefore, the particles cannot cross the energy barrier only by Brownian motion. In order to make coagulation take place, electrolytes should be added into the system to lower the height of the energy barrier so that the particles can cross it, causing coagulation. In other words, increasing the concentration or charge number of electrolytes will lead to the increase in κ value and the decrease in the thickness of the double layer $(1/\kappa)$, which will shorten the distance at which the overlapping of the double layer takes place. At this smaller distance, the van der Waals attraction will be relatively higher, thus leading to coagulation.

7.5.4 Critical Coagulation Concentration

It is well known that for a particular salt, there is a fairly sharply defined concentration at which coagulation will take place. This concentration can be referred to as the critical coagulation concentration (CCC). One of the earliest (1900) generalizations about the effect of electrolytes added is a result known as the Schulze–Hardy rule. This rule states that it is the valence of the ion of opposite charge to the colloid that has the principal effect on the stability of the colloid. The CCC value for a particular electrolyte is essentially determined by the valence of the counterion regardless of the nature of the ion with the same charge as the surface (ie, coion). It was found that the ratio of CCC values ($mol\ L^{-1}$) for mono-, di-, trivalent ions approximately equals to 1: $(1/2)^6$:$(1/3)^6$; that is, CCC values are inversely proportional to the sixth power of valence of counterion.

Now let us see how this result is understood in terms of the DLVO theory. As discussed previously, the stability of colloidal systems depends on the height of the energy barrier of the total potential energy curve. The height of the energy barrier decreases with the increase in the concentration of electrolytes. When the concentration of electrolytes reaches up to a certain value, the height of the energy barrier will equal to zero, as shown in Fig. 7.27, which is termed the critical coagulation state.

In this case, the concentration of electrolytes is the CCC for this colloidal system. For such a curve, the conditions $V = 0$ and $dV/dH = 0$ hold for the same value of H, that is,

$$V_T = V_R + V_A = 0 \tag{7.73}$$

and

$$\frac{dV_T}{dH} = \frac{dV_R}{dH} + \frac{dV_A}{dH} = 0 \tag{7.74}$$

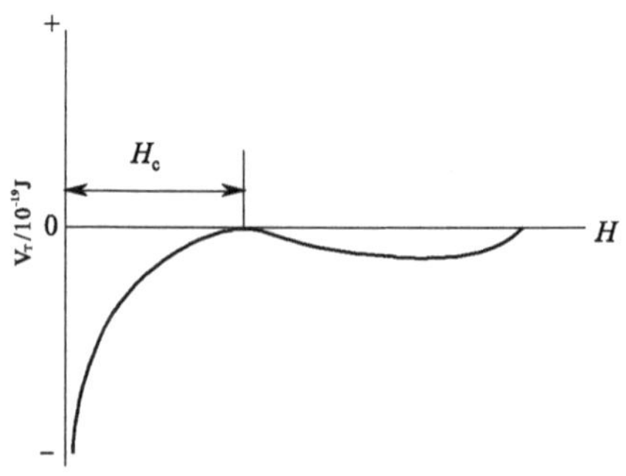

FIGURE 7.27 Total potential energy curve at critical coagulation state.

According to Eq. (7.58) and Eq. (7.72)

$$\frac{64n_0K_BT}{\kappa^2}\pi a\gamma_0^2e^{-\kappa H_c} - \frac{Aa}{12H_c} \tag{7.75}$$

and

$$\frac{64n_0K_BT}{\kappa}\pi a\gamma_0^2e^{-\kappa H_c} - \frac{Aa}{12H_c^2} = 0 \tag{7.76}$$

where H_c is the distance at which the system reaches the critical coagulation state. From Eq. (7.75) and Eq. (7.76), we obtain

$$\kappa H_C = 1$$

Substituting 1 for κH_C and Eq. (7.15) for κ in Eq. (7.76) and solving it gives n_0, which is the concentration of the cation or the anion of electrolyte at the critical coagulation state; therefore, we have the CCC as follows.

$$M = C\frac{\varepsilon^3(K_BT)^5\gamma_0^4}{A^2Z^6} \tag{7.77}$$

where c is a constant. As discussed previously, this result is obtained for spherical particles. The similar result can also be obtained for planar particles. A number of features of the DLVO theory emerge from this expression as follows:

1. At high potentials, γ_0 approaches 1 and the theory predicts that the coagulation concentrations of indifferent electrolytes containing counterions are inversely proportional to the sixth power of charge of counterions, which is the Schulze–Hardy rule obtained from the experiment. At very low potential, since $\gamma_0 = ze\psi_d/4K_BT$, the coagulation concentrations of indifferent electrolytes containing counterions are proportional to ψ_d^4/Z^2. Generally speaking, the coagulation concentration varies with the charge number of counterions in the range of Z^{-2} to Z^{-6}, which is in agreement with experimental evidence approximately.
2. From a typical experimental hydrosole coagulation concentration of 0.1 mol L^{-1} for $Z = 1$ (taking $\gamma_0 = 1$, $T = 298$ K, and $\varepsilon/\varepsilon_0 = 80$), the effective Hamaker constant A is calculated to be c. 2×10^{-9} J. Although slightly high, this value for A is of the order of magnitude expected from the theory of London–van der Waals forces. Considering the simplified nature of the model which has been used, the agreement between predicted and experimental coagulation concentrations is very good.
3. Coagulation concentrations for spherical particles of a given material at constant ψ_0 should be proportional to ε^3 and independent of the particle size. But the latter is obtained when the height of the energy barrier is assumed to be zero. Actually, particles typically have a certain kinetic energy; therefore,

they are able to cross the energy barrier with a certain height, leading to coagulation. Although the kinetic energy of particles is independent of the size of the particles, the height of the potential barrier is proportional to the size of the particles. Therefore, if the height of the energy barrier is not assumed to be zero, while it is assumed to be smaller than some value (such as 5 K_BT) as the condition for critical coagulation, large particles will be more stable than small particles.

According to the Schulze–Hardy rule or Eq. (7.77), the metal ions of higher valence should have lower coagulation concentration, that is, high coagulation efficiency; therefore, it is easy for us to understand that the coagulants commonly used in water treatment are aluminum (III), iron (II), and iron (III) inorganic salts, such as aluminum sulfate, aluminum chloride, ferric sulfate, ferric chloride, and ferrous sulfate. In addition, to form hydrous metal ions with charges of +3, they are able to form polynuclear hydroxo complex ions of higher charges by hydroxyl bridges, for example:

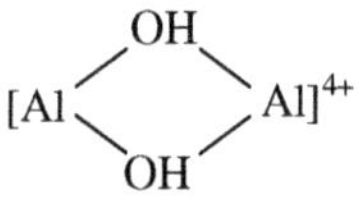

(7.78)

This dimmer may undergo further hydrolytic reactions to obtain more hydroxo groups to form more bridges and larger polynuclear hydroxo complex ions. A series of reactions have been reported for aluminum (III) ion in water:

$$2Al^{3+} + 2H_2O \rightleftharpoons Al_2(OH)_2^{4+} + 2H^+ \quad (7.79)$$

$$6Al^{3+} + 15H_2O \rightleftharpoons Al_6(OH)_{15}^{3+} + 15H^+ \quad (7.80)$$

$$7Al^{3+} + 17H_2O \rightleftharpoons Al_7(OH)_{17}^{4+} + 17H^+ \quad (7.81)$$

$$8Al^{3+} + 20H_2O \rightleftharpoons Al_8(OH)_{20}^{4+} + 20H^+ \quad (7.82)$$

$$13Al^{3+} + 34H_2O \rightleftharpoons Al_{13}(OH)_{34}^{5+} + 34H^+ \quad (7.83)$$

One of the most important reasons that aluminum (III) and iron (III) inorganic salts have high coagulation efficiency is the formation of such ions with higher charges. Based on this principle, the polyaluminum and polyfarric coagulants were developed in 1960s. Now they are applied in water treatment all over the world.

Commonly, coagulation takes place when the potential barrier equals zero or is very low as discussed previously. In this case, more particles are able to overcome the potential barrier by their kinetic energy, thus across the potential barrier. If the potential barrier is higher compared with the kinetic energy of the particles, the system should be stable. But if the secondary minimum is moderately deep compared with the kinetic energy, it should give rise to a

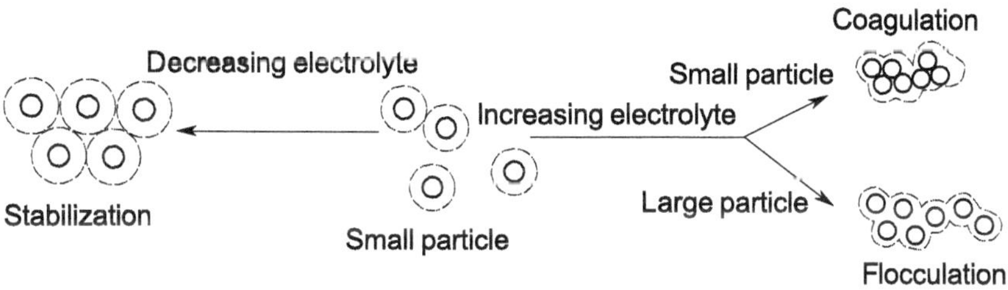

FIGURE 7.28 **Effect of electrolyte and particle size on the stability of colloidal systems.**

loose, easily reversible flocculation (the aggregation at the secondary minimum is termed flocculation). For small particles, the secondary minimum is never deep enough for this to happen in those cases where the potential barrier is high enough to prevent normal coagulation into the primary minimum. If the particles are larger, flocculation in the secondary minimum may cause observable effects. Fig. 7.28 illustrates the effect of electrolytes and particle size on the stability of the colloidal system.

Example 7.5 Coagulation Concentration

The coagulation concentrations of $NaNO_3$, $Mg(NO_3)_2$, and $Al(NO)_3$ for aqueous sole of AgI are 140, 2.60, and 0.067 mol dm^{-3}, respectively. Indicate if the sole particles are negatively charged or positively charged.

Solution

All of the anions of these three electrolytes are the same, NO_3^-, while the cations of these three electrolytes are different, Na^+, Mg^{2+}, and Al^{3+}, respectively. Suppose the particles are positively charged, the counterion should be the same and the coagulation concentration should be approximately equal according to the Schulze–Hardy rule, but this is not in accord with that given in this example. Suppose the particles are negatively charged, the counterion should be different and there must be remarkable differences between the three coagulation concentrations according to the Schulze–Hardy rule, the ratio of which should approximately be equal to 100:1.6:0.14, which is approximately in accordance with that given in this example. Therefore, the particles must be negatively charged.

Example 7.6 Coagulation Concentration

For the coagulation of a positively charged aqueous sol, the CCC of KNO_3 is 50×10^{-3} mol dm^{-3}. Evaluate the CCC of K_2SO_4 for this aqueous sol.

Solution

According to the Schulze–Hardy rule

$$\frac{x}{50\times10^{-3}}=\left[\frac{Z(NO_3^-)}{Z(SO_4^{2-})}\right]^6=\left(\frac{1}{2}\right)^6$$

where x is the CCC of K_2SO_4 and Z is the charge number of the ion. Solving it gives

$$X = 0.78 \times 10^{-3} (\text{mol dm}^{-3})$$

7.6 KINETICS OF COAGULATION

Only the coagulation with a certain rate can meet the demand for the amount of water produced in a water plant, thus, it has a practical meaning. It is known from DLVO theory that the colloidal system is stable due to the existence of potential barrier in the total potential energy curve. If the potential barrier equals zero, every collision of particles will cause coagulation, which is termed rapid coagulation. However, in most practical cases of interest we are concerned with the reduction in collision rate (and the consequent increase in stability) caused by potential barriers introduced through interparticle repulsive forces, which is termed slow coagulation. Commonly, particles in suspension collide with each other as a consequence of two mechanisms of particle transport:

1. Particles move because of their thermal energy (Brownian motion). The coagulation resulting from this mode of transport is referred to as perikinetic coagulation.
2. If particles are sufficiently large or the fluid shear rate is high, the relative motion from velocity gradients can exceed that caused by the Brownian effect. The coagulation resulting from this mode of transport is referred to as orthokinetic coagulation.

7.6.1 Rate of Perikinetic Coagulation

7.6.1.1 Rapid Coagulation

As introduced previously, in perikinetic coagulation the collisions of particles are caused by their Brownian motion, thus the frequency of collision depends on the thermal diffusion movement of particles. As early as 1917, Smoluchowski applied diffusion theory to coagulation and formulated the equations of the collision rate for particles transported by diffusion alone.

Here we begin by considering a stationary spherical particle of radius R_0 as a reference sphere. Suppose the other spherical particle of the same radius R_0 at the distance r from the center of the reference sphere is moving toward it, the distance between them will gradually decrease while they approach each other. When r reaches $2R_0$, they will collide with each other. Let us make a spherical surface of the radius of $2R_0$ around the center of the reference sphere, then the rate at which the other particles collide with the reference sphere equals the rate at which they diffuse across this spherical surface and can be obtained from Fick's first law.

$$J = -DA\frac{dc}{dx}$$

Substituting the area of spherical surface for A and considering $x = -r$ (the direction of diffusion is opposite to that of distance from the centre of reference sphere) gives

$$J = D \cdot 4\pi (2R_0)^2 \left(\frac{dc}{dr} \right)_{r=2R_0} \tag{7.84}$$

Since the number of particles entering any spherical surface around the same sphere center at the same time at stable state is equal, we have

$$J = D \cdot 4\pi r^2 \frac{dc}{dr} \tag{7.85}$$

thus

$$dc = \frac{J}{D \cdot 4\pi r^2} dr \tag{7.86}$$

Integrating

$$\int_0^{c_0} dc = \int_{2R_0}^{\infty} \frac{J}{D \cdot 4\pi r^2} dr$$

thus

$$c_0 = \frac{J}{8\pi D R_0} \tag{7.87}$$

That is,

$$J = 8\pi R_0 D c_0 \tag{7.88}$$

Actually, the reference sphere is also making the Brownian motion at the same time, therefore, Eq. (7.88) should be doubled

$$J = 16\pi R_0 D c_0 \tag{7.89}$$

thus for all particles, the rate at which these particles disappear is given by

$$-\frac{dc}{dt} = c_0 \cdot J = 16\pi R_0 D c_0^2 \tag{7.90}$$

It is noted that from Eq. (7.89) to Eq. (7.90), actually, we calculate the collision for a pair of particles two times; therefore, the rate at which these particles disappear should be

$$-\frac{dc}{dt} = 8\pi R_0 D c_0^2 \tag{7.91}$$

Eq. (7.91) indicates that the reaction is a second order one. Integrating this equation gives

$$c = \frac{c_0}{1 + 8\pi R_0 D c_0 t} \tag{7.92}$$

where c is the concentration of particles of the initial size after the reaction for time t.

Example 7.7 Rate of Perikinetic Coagulation

The perikinetic coagulation of a golden sole takes place. Suppose the potential energy barrier is zero, the temperature T 291 K, the viscosity η 1.06 × 10^{-3} Pa s, the initial concentration 5.22 × 10^{14} m^{-3}, and the radius of particles 9.6×10^{-8} m. Calculate the concentration of particles after the reaction for 900 s.

Solution

Combining Eq. (2.8) and Eq. (2.9) gives

$$D = \frac{K_B T}{6\pi\eta R_0} = \frac{1.38 \times 10^{-23} \times 291}{6 \times 3.14 \times 1.06 \times 10^{-3} \times 9.0 \times 10^{-8}} = 2.23 \times 10^{-12}$$

After the reaction for 900 s, the concentration of particles of the initial size is

$$c = \frac{c_0}{1 + 8\pi R_0 D c_0 t} = \frac{5.22 \times 10^{14}}{1 + 8 \times 3.14 \times 9.6 \times 10^{-8} \times 2.23 \times 10^{-12} \times 5.22 \times 10^{14} \times 900}$$
$$= 1.42 \times 10^{14} (\text{m}^{-3})$$

7.6.1.2 *Slow Coagulation*

Fuchs assumed that the effect of the potential barrier corresponds to the repulsion between particles, under which the particles go away from each other. Since the existence of the potential barrier, the Smoluchowski equation, which shows the diffusion of particles toward the reference spherical particle, should be modified as

$$J = D \cdot 4\pi r^2 \frac{dc}{dr} + c \tag{7.93}$$

where c is the correction factor for resistance, which is the number of particles leaving the reference sphere at unit time. Considering the relation between the correction factor for resistance and the interaction potential energy, as well as the Brownian motion of the reference spherical particle, we obtained the rate equation for slow coagulation

$$J = \frac{8\pi D c_0}{\int_{2R_0}^{\infty} \exp\left(\frac{V_T}{K_B T}\right) r^{-2} dr} \tag{7.94}$$

Generally, W is known as the stability ratio, that is,

$$W = \frac{K_r}{K_s} \tag{7.95}$$

where K_r and K_s are the rate constants for rapid coagulation and slow coagulation, respectively. The stability ratio is an important factor, which shows the stability degree of the dispersion system. The smaller the K_s, the greater the W, and the more stable the system. If $W = 1$, the coagulation will become rapid coagulation, and the system will be unstable. Comparing Eq. (7.94) with Eq. (7.88) gives

$$W = \frac{K_r}{K_s} = 2R_0 \int_{2R_0}^{\infty} \exp\left(\frac{V_T}{K_B T}\right) r^{-2} dr \tag{7.96}$$

or

$$W = \frac{K_r}{K_s} = 2 \int_{2}^{\infty} \exp\left(\frac{V_T}{K_B T}\right) S^{-2} ds \tag{7.97}$$

where

$$s = \frac{r}{R_0}$$

Eq. (7.97) shows that the value of W depends on the interaction of the potential energy of particles. Theoretical relationships between the stability ratio and electrolyte concentration can be obtained by numerical solution of this integral for the given values of A and ψ_d. Fig. 7.29 shows the experimental result for AgI sol (52 nm), which is in accordance with the result of the calculation. For constant ψ_d, a linear relationship between log W and log c is predicted for the whole of the slow coagulation region. The turning point indicates the CCC.

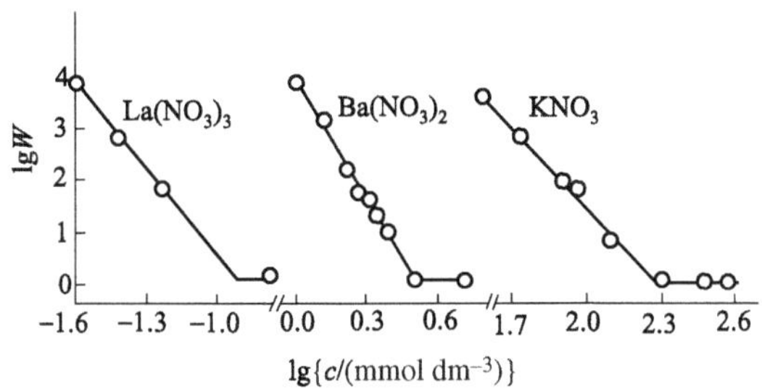

FIGURE 7.29 Experimental result of lg$W$–lg$c$ for AgI sol (52 nm).

7.6.2 Rate of Orthokinetic Coagulation

The perikinetic coagulation caused by the Brownian motion is too slow to apply alone to water treatment in practice, especially when particles become larger due to aggregation; the Brownian motion will be weak, even stop, thus the coagulation will become weak or stop. But it has been found for a long time that slow stir is capable of accelerating coagulation because it can produce the shearing strain with the velocity gradient, leading to the relative movement between particles, thus causing the collision of particles.

For a shearing strain with uniform velocity gradient, the simple theory of coagulation kinetics has been developed. But it is difficult to find such a shearing strain filed with uniform velocity gradient in practical situations. Most of the situations are turbulent flow filed. In the past decades the theory of uniform velocity gradient was inadequately extended to turbulent flow, which has no uniform velocity gradient, but is very common in practice. Here let us discuss the coagulation kinetics of these two situations respectively.

7.6.2.1 Uniform Shearing Strain Field

As far as the coagulation in uniform shearing strain field is concerned, the work of Smoluchowski is the cornerstone of coagulation. First we discuss rapid coagulation. Because it is relative movement that causes the collision of particles, let us again assume one particle as a stationary reference sphere (j) in medium, as shown in Fig. 7.30.

If the shearing strain field is not disturbed by particles, the pathway of particles will be linear. In rapid coagulation every collision causes aggregation just as in the rapid coagulation of perikinetic coagulation. Suppose a particle i above the center line moves in the direction x. When its center locates in the range of upper hemi-column radius $R_{ij}(r_i + r_j)$, it will collide with particle j. The velocity

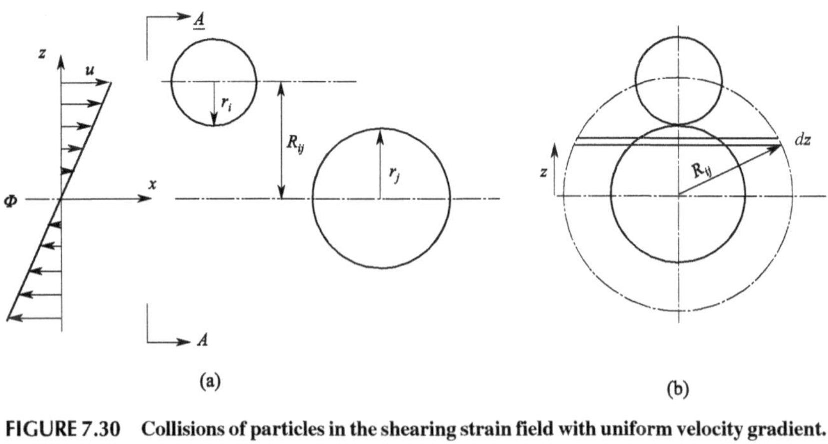

FIGURE 7.30 Collisions of particles in the shearing strain field with uniform velocity gradient. (a) Relative movement and (b) collision with each other.

of particle i relative to particle j depends on the distance from the plane x. Let the distance be Z and the velocity gradient dv/dz, then the relative velocity of particle i can be expressed as $Z(dv/dz)$. The flow rate of fluid outflowing from the upper hemi-column is the flow rate of fluid passing through the cross section of the upper hemi-column, as shown in Fig. 7.28b, and the area of cross section with the height of dz is expressed as $2(R_{ij}^2 - Z^2)^{1/2}dz$. Thus the amount of fluid passing through this micro cross section per unit time is

$$dQ = 2(R_{ij}^2 - Z^2)^{1/2}dz \cdot Z(dv/dz) \tag{7.98}$$

Therefore, the amount of fluid passing through the cross section of the upper hemi-column per unit time is

$$Q_{1/2} = 2(dv/dz)\int_0^{R_{ij}} Z\left(R_{ij}^2 - z^2\right)^{1/2} dz \tag{7.99}$$

Similarly, there is a flow in the direction opposite to x in the lower hemi-column; therefore, the total flow rate relative to particle j is

$$Q = 4(dv/dz)\int_0^{R_{ij}} Z\left(R_{ij}^2 - z^2\right)^{1/2} dz \tag{7.100}$$

Integral of Eq. (7.100) gives

$$Q = \frac{4}{3}(dv/dz)R_{ij}^3 \tag{7.101}$$

Since the number of particle i in unit volume of fluid is N_i, the rate of the collision of particle i with particle j is

$$J = \frac{4}{3}N_i(dv/dz)R_{ij}^3 \tag{7.102}$$

If the number of particle j in unit volume of fluid is N_j, the total rate of the collisions of particles with each other will be

$$J = N_i N_j (dv/dz)\ R_{ij}^3 \tag{7.103}$$

Just as in perikinetic coagulation, the concentration of particles of size $k(=i+j)$ increases due to the collisions of particles i with particles j and decreases due to the collisions of particles k with other particles. Therefore, the change rate of concentration of particle k can be expressed as

$$\frac{dN_k}{dt} = \frac{1}{2}\sum_{i=1}^{i=k-1}\frac{4}{3}N_i N_j (dv/dz)R_{ij}^3 - N_k \sum_{i=1}^{\infty}\frac{4}{3}N_i (dv/dz)R_{ik}^3 \tag{7.104}$$

where N is the number concentration of the particles; k, i, and j are the subscripts denoting particle size; and t is the reaction time. It is noted that from Eq. (7.102) to Eq. (7.103), actually, we calculate the collision for a pair of particles two times; therefore, "1/2" is added before the first item in Eq. (7.104).

Relate R_{ij} to r_1 (radius of primary particle) and suppose the particle aggregation is the simple combination of particles, that is, the volume x_i of particle i equals to i times the volume x_1 of primary particles, then we have

$$X_i = i\frac{4}{3}\pi r_1^3 = \frac{4}{3}\pi r_i^3$$

thus

$$ir_1^3 = r_i^3$$

Since

$$R_{ij} = r_i + r_j$$

then

$$R_{ij}^3 = r_1^3\left(i^{\frac{1}{3}} + j^{\frac{1}{3}}\right)^3 \tag{7.105}$$

Applying Eq. (7.105) and substituting G for dv/dz, and considering all particles with different size, Eq. (7.104) becomes

$$\frac{d\sum_{k=1}^{\infty} N_k}{dt} = \frac{2Gr_1^3}{3}\left[\sum_{i=1}^{\infty}\sum_{j=1}^{\infty} N_i N_j\left(i^{\frac{1}{3}} + j^{\frac{1}{3}}\right)^3 - 2\sum_{i=1}^{\infty}\sum_{k=1}^{\infty} N_i N_k\left(i^{\frac{1}{3}} + k^{\frac{1}{3}}\right)^3\right] \tag{7.106}$$

Since all possible combinations of collision are considered, particles k include particles i and j. If $i \approx j \approx k$, we will have

$$\frac{d\sum_{k=1}^{\infty} N_k}{dt} = \frac{16Gr_1^3}{3}\left(\sum_{k=1}^{\infty} N_k^2 k - 2\sum_{k=1}^{\infty} N_k^2 k\right) = -\frac{16Gr_1^3}{3}\sum_{k=1}^{\infty} N_k^2 k \tag{7.107}$$

where minus indicates that the number of particles always decreases during the coagulation process. Suppose the system is a monodispersed system at the initial stage of reaction, and $k = 1$ and $N_k = N_0$ when $t = 0$, the previous equation will be

$$-\frac{dN_1}{dt} = \frac{16}{3}Gr_1^3 N_0^2$$

or

$$-\frac{dN}{dt} = \frac{16}{3} G r^3 N^2 \tag{7.108}$$

which indicates that coagulation is the second order reaction with respect to the number concentration of particles. The total volume of N particles is constant and given by

$$\varphi = N\left(\frac{4}{3}\pi r^3\right) \tag{7.109}$$

where φ is the total volume of particles per unit volume of the system and therefore is dimensionless. Substituting Eq. (7.109) into Eq. (7.108) gives an equation of the first order reaction as follows:

$$-\frac{dN}{dt} = \frac{4}{\pi}\phi G N \tag{7.110}$$

Integral of Eq. (7.110) gives

$$-\ln\frac{N}{N_0} = \frac{4}{\pi}\phi G t \tag{7.111}$$

and the half-life period is given by

$$t_{1/2} = \frac{0.693}{\frac{4}{\pi}\phi G} \tag{7.112}$$

From Eq. (7.112), it is concluded that

1. Increasing velocity gradient will shorten the half-life period. But the increase in velocity gradient is limited in practice; therefore, its effect is not very high.
2. Combining with Eq. (7.109) gives

$$\frac{t_{1/2(\text{large})}}{t_{1/2(\text{small})}} = \frac{\phi_{\text{small}}}{\phi_{\text{large}}} = \left(\frac{r_{\text{small}}}{r_{\text{large}}}\right)^3 \tag{7.113}$$

This implies that the half-life period of particles of radius of 10 μm is only one-thousandth that of particles of radius of 1 μm, and the half-life period will quickly shorten with the continued increase in particle volume during the coagulation process. It is very important to know that the rate of coagulation must be very high when some large particles exist in the beginning of the stir.

In rapid coagulation theory, it is assumed that every collision will cause aggregation, that is, all particles are destabilized. Actually, only some of the particles in the system are destabilized, that is, only a part of the collisions is efficient. This part of collisions can be shown in terms of a factor α (collision efficiency). When $\alpha = 1$, the coagulation is termed rapid coagulation, while $\alpha < 1$, the coagulation is termed slow coagulation. Thus the equation for the rate of orthokinetic coagulation (7.108) becomes

$$-\frac{dN}{dt} = \frac{16}{3}\alpha G r^3 N^2 \tag{7.114}$$

O'Melia and Stumm said that coagulation in natural systems and in water and wastewater treatment systems depends on the variable of Eqs. (7.114) or (7.111). In natural waters long detention times may provide sufficient contact opportunities despite very small collision frequencies (small G and small ϕ). In freshwater the collision efficiency is usually also low ($\alpha \sim 10^{-4}$–10^{-6}, that is, only 1 out of 10^4–10^6 collisions leads to a successful agglomeration). In seawater, colloids are less stable ($\alpha \sim 0.1$–1) because the salinity compresses their double layer. Estuaries with their salinity gradients and tidal movements represent gigantic natural coagulation tanks where much of the dispersed colloidal matter of the rivers becomes settled. In water and wastewater treatment systems we can reduce detention time (volume of tank) by adding coagulants at the proper dosage ($\alpha \rightarrow 1$) and by adjusting power input (G). If the concentration φ is too small, it can be increased by adding additional colloids as so-called coagulation aids.

7.6.2.2 Nonuniform Shearing Strain Field

In order to use the aforementioned Smoluchowski equation of laminar coagulation in turbulent coagulation condition, Camp and Stein developed G-value as follows:

$$G = \left(\frac{\varepsilon}{\nu}\right)^{1/2} \tag{7.115}$$

where ε is the power input per unit mass, and ν is the dynamic viscosity of fluid. From that time, it has been widely accepted that the mixing conditions with the same G-value are almost identical to the coagulation viewpoint, and the G-value has been widely used in the design and scale-up of mixing process components. But since the middle of 1980s, some research results have been published that criticize and question the use of the G-value as a valid basis for the design of coagulation basins; for example, Cleasby said that the G-value is only valid for vortices whose size is smaller than those necessary for coagulation in both water and wastewater treatment. Han and Lawler stated that the importance of the velocity gradient has been apparently overemphasized in the traditional view of coagulation. In addition, it was proved that the grid flocculation developed in recent years exhibits excellent performance, but the G-value equals zero at some distances from the grids. Therefore, many researchers attempt to discuss the coagulation dynamics in turbulent condition directly from turbulent flow theory in the last several decades. According to Kolmogoroff's local isotropic theory, the energy input into fluid is primarily used for the formation of large vortices. These large vortices accomplish most of the momentum transport but only a small amount of energy dissipation. The energy transfer in turbulent flow results in the formation of turbulent vortices of gradually decreasing size until the energy is dissipated by viscous forces; that is, there are series transformed from large vortices to small vortices in turbulent flow. But only the Kolmogoroff's microscale vortices, which have the same size approximately as the particles, are thought to be responsible for flocculation. Kolmogoroff's microscale can be computed from

$$\eta = \left(\frac{\nu^3}{\varepsilon}\right)^{1/4} \tag{7.116}$$

But this microscale vortices theory is only applicable to microscale vortices and does not concern vortices of other sizes. In summary, the reasons that induce the turbulent flocculation are still not very clear or complete up to now.

7.7 EFFECT OF MACROMOLECULES ON COLLOID STABILITY

7.7.1 Stabilization Effect of Macromolecules

It is well known that some colloidal particles can be protected by the macromolecular substances adsorbed at their surface. But these phenomena cannot be well explained by the increase in electric repulsion or the decrease in van der Walls attraction. The mechanism of protection is generally complex and a number of factors may be involved.

7.7.1.1 Effect of Macromolecular Structures

Generally, the protective macromolecule consists of a stabilizing group and an adsorbing group in one molecule. Since the adsorbing group has the stronger affinity for the surface of colloidal particles, the macromolecules can adsorb onto the particle; since the stabilizing group has the stronger affinity for water, the particles are dispersed in water medium. There are two kinds of protective macromolecules of different structures: graft copolymer and segmented copolymer. Fig. 7.31 illustrates the mechanism of the stabilizing effect of macromolecules for colloidal particles.

7.7.1.2 Effect of Molecular Weight and Concentration of Macromolecules

The higher the molecular weight of macromolecules, the thicker the adsorption layer and the better the stability effect. If the concentration is very small, the bridge flocculation will take place due to the existence of empty surfaces on the

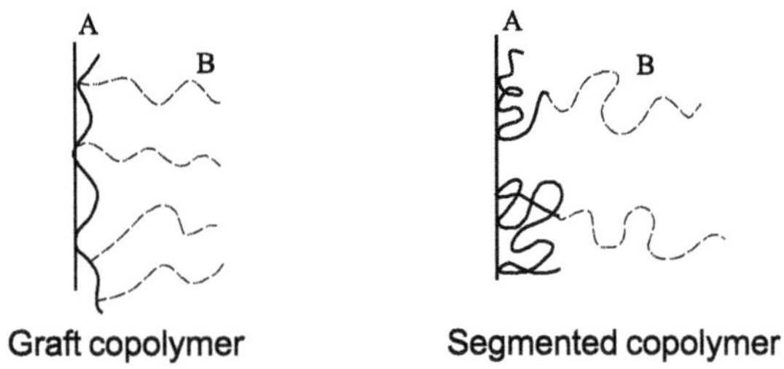

FIGURE 7.31 Stabilizing effect of macromolecules for colloidal particles. A, Adsorbing group; and B, stabilizing group.

particles. But there is no need for very high concentration because it is useless; therefore, the concentration should be suitable, not very high and not very low, to stabilize the system.

7.7.1.3 Effect of Medium Property

If the medium is a good solvent for the stabilizing group of macromolecules, the macromolecule will have a stabilizing effect for the system. If such a poor solvent is added to the system that the solvent becomes θ, the system will become destabilized and start to aggregate. Similarly, if the temperature is changed to θ, the system will also become destabilized and aggregation will take place.

The mechanism of stabilization of a system is explained as follows.

7.7.1.4 Effect of Space Restriction

Suppose the particles with the adsorbed macromolecules approach each other; the macromolecule layers will be compressed, as shown in Fig, 7.32a, thus leading to the decrease in the space occupied by them and the decrease in arranged configurations of molecules, which implies the decrease in entropy and increase in free energy, thus causing the repulsion between particles. This effect is referred to as the effect of space restriction.

7.7.1.5 Effect of Mixing

Suppose the particles with the adsorbed macromolecules approach one another; the adsorbed macromolecule layers will permeate each other, as shown in

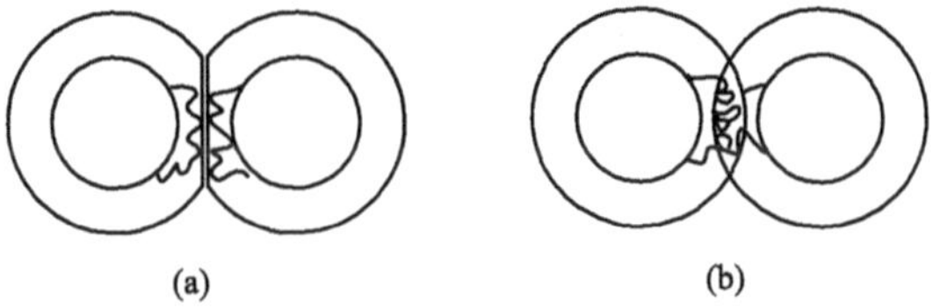

FIGURE 7.32 Effect of space restriction and mixing. (a) Effect of space restriction (compress, not permeate) and (b) effect of mixing (permeate, not compress).

Fig. 7.32b, leading to the increase of the concentration of chain segments between two particles, thus causing the increase in attraction or repulsion, which depends on the interaction between polymer and medium. This is referred to as the effect of mixing.

If the interaction of particles is repulsive, there must be an expression of thermodynamics as follows.

$$\Delta G_{\mathrm{R}} = \Delta H_{\mathrm{R}} - T\Delta S_{\mathrm{R}} > 0 \tag{7.117}$$

Thus two mechanisms of stabilization are proposed:

1. $\Delta H_{\mathrm{R}} > 0, \Delta S_{\mathrm{R}} > 0$, and $\Delta H_{\mathrm{R}} > T\Delta S_{\mathrm{R}}$: enthalpy stabilization;
2. $\Delta H_{\mathrm{R}} < 0, \Delta S_{\mathrm{R}} < 0$, and $\Delta H_{\mathrm{R}} < |T\Delta S_{\mathrm{R}}|$: entropy stabilization.

7.7.2 Application of Stabilization Effect of Macromolecules in Cooling Water

Stabilizing effects of macromolecules is used in the treatment of circulating cooling water in industry. Currently, the open type of circulating cooling water is wildly applied in many kinds of industries, including chemical and electrical works. The serious scaling takes place in this type of system due to the supersaturation of salt caused by evaporation and concentration. In order to avoid or decrease the scaling, many kinds of macromolecules, for example, polymaleic acid, polyacrylic acid, and their copolymers with other monomers, are added in the cooling water. They are called macromolecular inhibitors, which make the colloidal particles, such as calcium carbonate, magnesium hydroxide, calcium phosphate and silicate, etc., dispersed and suspended in water due to the stabilization effect when they are added, thus decreasing the scaling in the system of cooling water.

7.7.3 Flocculation Effect of Macromolecules

In the early years, it was found that colloidal dispersions can be made more sensitive to precipitation by the addition of small quantities of macromolecules. These macromolecules are referred to as flocculants currently. The macromolecular flocculants commonly used in early times are polyelectrolytes. Their flocculation mechanism is considered to be the simple charge neutralization. If the charges of macromolecular ions are opposite to that of particles, flocculation and precipitation will take place. But in the later period, it has been discovered that some nonionic macromolecules, even the macromolecules with the same charge as the colloidal particles, have sensitive or flocculation effect. Therefore, neutralization is not the only mechanism of their flocculation. Now we think long-chain additives, such as gelatin, can sometimes bring about a rather loose flocculation by a bridging mechanism in which the molecules are adsorbed with part of their chain on two or more particles. Such flocculation normally occurs

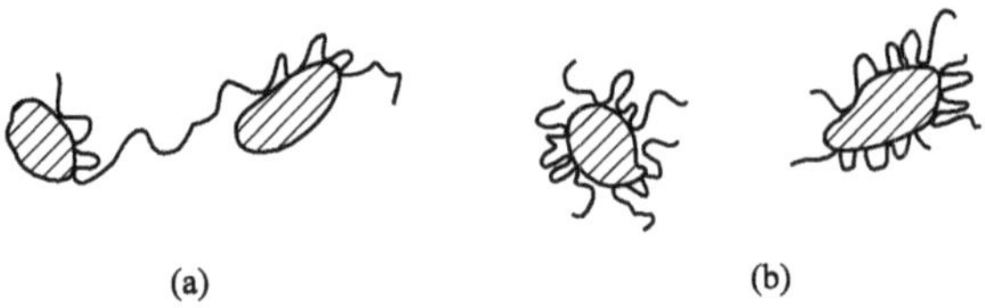

FIGURE 7.33 Principle of bridging flocculation and protective effect of macromolecule. (a) Flocculation and (b) protection.

over a narrow range of additive concentrations; at a higher concentration protective action is obtained, since bridging can occur only under conditions in which the remaining particles still have empty surfaces, thus further adsorption of the additive is possible. Fig. 7.33 illustrates the principle of bridging flocculation and the protective effect of macromolecules.

According to the principle of bridging flocculation, the molecule of macromolecular flocculant must be adsorbed on both of the particles at the same time to bridge between them; therefore, the macromolecular flocculants commonly used are homopolymers. The efficiency of bridging flocculation is influenced by the molecular weight and charge density of macromolecules. Generally speaking, high molecular weight is favorable for bridging, but too high molecular weight will produce the overlap of chain segments and greater repulsion between them, thus weakening bridging effect and leading to low flocculation efficiency. High charge density is helpful for molecules to extend in solution, thus favorable for bridging, but if the sign of the charge is the same as that of the particle, too high charge density will weaken the adsorption of the chain segment on the particles, thus leading to poor bridging and low flocculation efficiency. Therefore, there should be an appropriate hydrolysis degree for polyacromide, which is the most popular flocculant currently in water treatment. Commonly, the hydrolysis degree of polyacromide is controlled to reach 30%.

Many natural macromolecule products can be used as flocculants in water treatment, such as starch, protein, cellulose, chitosan, and their various modified products. In recent decades, the artificial macromolecular flocculants have been developed greatly. Currently, the commonly used artificial flocculants in water treatment include polyacromide (PAM), polydimethyldiallylammonium (PDMDAAC), polyethyleneimine (PEI), and their various modified products.

7.8 COAGULATION IN NATURAL WATERS AND WATER TREATMENT

Coagulation is a process that makes finely divided particles aggregate and form large flocs, which can be settled and separated from water.

Coagulation is of great importance in the transport and distribution of suspended colloidal particles in natural waters. Although the flocs formed tend to be distributed by convective mass transfer, their distribution is also influenced by the force of gravity. Since many dissolved substances adsorb onto colloids or react chemically with colloid surface groups, coagulation also plays a role in the transformation of solutes. Oxide, especially those of Si, Al, and Fe, are abundant components of the earth's crust. They participate in many chemical processes in natural water and often occur as colloids in water. The colloid stability of hydrous oxides may be affected by electrolytes in a different way than that of hydrophobic colloids. Specific adsorption of cations and anions on hydrous oxide surfaces may be interpreted as surface coordination reaction. As indicated in Eqs. (7.118)–(7.121), cations and anions typically present in natural water become specifically adsorbed at metal oxide surface and modify its charge.

$$\equiv \mathrm{MeOH} + \mathrm{M}^{z+} \rightleftharpoons \equiv \mathrm{MeOM}^{(z-1)} + \mathrm{H}^{+} \tag{7.118}$$

$$_{2}\equiv \mathrm{MeOH} + \mathrm{M}^{z+} \rightleftharpoons (\equiv \mathrm{MeO})_{2}\mathrm{M}^{(z-2)} + 2\mathrm{H}^{+} \tag{7.119}$$

$$\equiv \mathrm{Me}^{-}\mathrm{OH} + \mathrm{A}^{z-} \rightleftharpoons \equiv \mathrm{Me}^{-}\mathrm{A}^{(z-1)} + \mathrm{OH}^{-} \tag{7.120}$$

$$_{2}\equiv \mathrm{Me}^{-}\mathrm{OH} + \mathrm{A}^{z-} \rightleftharpoons (\equiv \mathrm{MeO})_{2}\mathrm{A}^{(z-2)} + {}_{2}\mathrm{OH}^{-} \tag{7.121}$$

Currently, coagulation operation is one of the most important units of water treatment and frequently is a necessary unit which cannot be omitted or substituted. The targets of coagulation are the colloidal and suspended particles which consist of insoluble and hydrophobic materials. Therefore, the attempt to remove soluble materials from water by coagulation directly is impracticable. If some soluble materials can be transformed to insoluble materials, they will be coagulated. In some cases, soluble materials can be removed to some extent since they are adsorbed onto the flocs formed in the coagulation process; as a result, the coprecipitation will take place, which can be considered to be a synergy effect. Generally, the flocs produced can be separated from water by settling, filtration, flotation, etc. The purpose of coagulation is not only to cause particle settling but also to give particles the appropriate surface property which facilitate particles to be retained by the filter medium in subsequent filtration unit. The aim of water treatment for the water supply is to provide drinking water for people and industrial water for production. In water plants for the water supply, surface water or underground water is treated to meet the standards of usage by people or the industry. The typical technological process includes natural sedimentation, coagulation, sedimentation, filtration, and disinfection. The treatment of industrial wastewater or domestic sewage is commonly classified in three stages, which are first-stage treatment, second-stage treatment, and third-stage treatment. The first-stage treatment (or primary treatment) consists of screening, gravity settling, and flotation, which are used consecutively to remove particles larger than 100 μm. After the primary treatment, the second-stage treatment (or secondary treatment) commonly consists of the biochemical method and the coagulation method. The former is designed to remove the residual organic substances, and the latter is designed to remove the residual suspended inorganic substances and the insoluble organic substances. After the primary and secondary treatment, the water can meet the water quality standards for agriculture irrigation and sewage discharge. But it still contains some trace pollutants, for example, suspended matters, persistent organic pollutants, soluble inorganic substances, algae nutrients containing nitrogen and phosphorus, virus and bacterium, etc. If discharged to small rivers, it will result in pollution due to the limited dilution. In such cases, the third-stage treatment (or tertiary treatment) is needed. In the tertiary treatment, we use many kinds of physical and chemical methods, including aeration, adsorption, coagulation, sedimentation, sand filtration, ion exchange, electrodialysis, and reverse osmosis to reach the high requirements. It can be concluded that among all of the aforementioned technological processes, coagulation is a very common method and very important unit that often cannot be omitted or substituted.

Since 1974, people have discovered that drinking water possibly contains volatile trihalomethanes (THMs) and some nonvolatile organic chlorides which are produced from natural organic matter (NOM) in the disinfection unit of water treatment and termed disinfection byproducts (DBPs). These DBPs may lead to mutation and cancer and therefore are very harmful to human health. In order to eliminate DBPs, USEPA, and AWWA proposed "enhanced coagulation" to remove the precursors of DBPs, that is, NOM. The enhanced coagulation is carried out by appropriate increase in coagulant dosage and the optimization of some operation conditions. The result shows that enhanced coagulation is the most practicable technology for eliminating DBPs compared with the other complicated and expensive equipment modifications.

Chapter 8

Surface of Liquid

Chapter Outline

8.1 SURFACE TENSION AND SURFACE FREE ENERGY

8.1.1 Basic Concepts

The separation area between two phases is termed the interface; if one of the two phases is gas, the interface is called the surface. Fig. 8.1 represents a loop of wire with one movable side on which a film could be formed by dipping the frame into a liquid. Let AB be the movable side with length L. If a force F is perpendicularly applied to the movable side to increase the area of film isothermally and reversibly, the operator will feel a force, which is of the same value as F but in the direction opposite to F. This opposite force tends to cause the slide wire to move in the direction of decreasing film area.

Colloid and Interface Chemistry for Water Quality Control. http://dx.doi.org/10.1016/B978-0-12-809315-3.00008-6

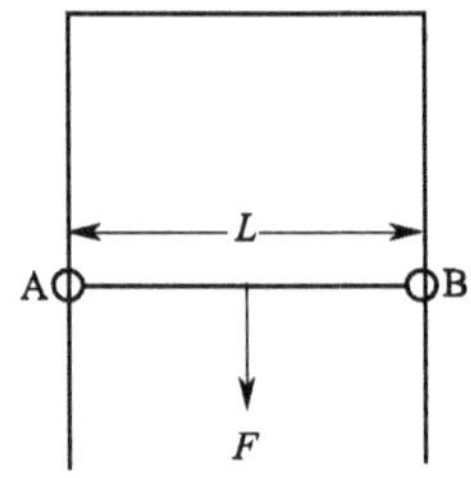

FIGURE 8.1 **Surface tension.**

The opposite force F evidently operates along the entire edge of the film and will vary with the length L of the slide wire. Since the film in Fig. 8.1 has two sides, F is given by

$$F = \gamma.2L \tag{8.1}$$

where γ is termed surface tension (or interface tension), which is the opposite force per unit length of edge. Eq. (8.1) defines the unit of surface tension to be nN m^{-1}. In summary, surface tension (or interface tension) is a force which operates on a surface and acts perpendicularly and inward from the boundaries of the surface, tending to decrease the area of the surface.

Application of a force infinitesimally larger than the equilibrium force to the slide wire in Fig. 8.1 will displace the wire through a distance dx. The product of force and distance equals energy; in this case the energy spent in increasing the area of film by the amount $dA = 2Ldx$, where A is the area of the film. Therefore the nonvolume work done on the system is given by

$$\delta W' = F \cdot dx = \gamma \cdot 2L \cdot dx = \gamma \cdot dA \tag{8.2}$$

This provides a second definition of surface tension: it equals the work per unit area required to produce a new surface. In terms of this definition, the units of γ are energy J m^{-2}.

It can be seen, therefore, that there are two equivalent interpretation of γ: force per unit length of boundary of the surface and energy per unit area of the surface, that is, the surface free energy. If the numerator and denominator of the former is multiplied by length, its dimension (force length^{-1}) will change to the dimension of the latter (energy area^{-1}), therefore, the dimensional equivalency of the two is evident.

8.1.2 Origination of Surface Tension and Surface Free Energy

Fig. 8.2 illustrates an air–liquid interface. The molecules which are located within the bulk of a liquid are, on average, subjected to equal forces of attraction in all directions, whereas those located at an air–liquid interface experience unbalanced attractive forces because the attractive force between gas molecules and liquid molecules is smaller compared with that between liquid molecules,

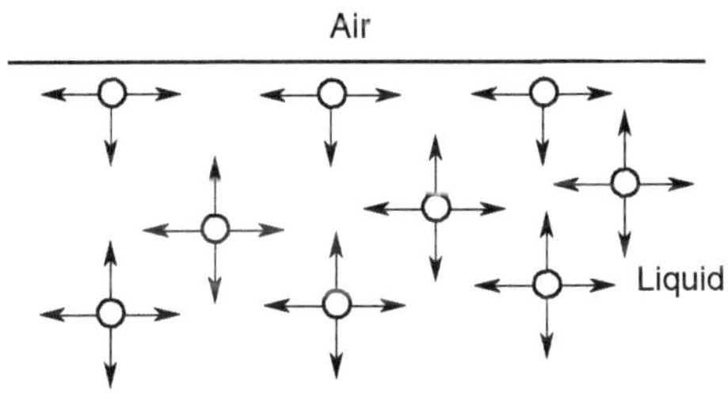

FIGURE 8.2 **Origination of surface tension.**

thus resulting in a net inward pull, that is, a vertically inward composite force. As a result, as many molecules as possible will leave the liquid surface for the interior of the liquid, and the surface will, therefore, tend to contract spontaneously, thus showing the surface tension.

To increase the surface area, we should migrate the molecules in the interior of the liquid to the air–liquid interface; therefore, the work must be done against the net inward pull. This work will transform to the free energy of the surface, and as a result, the energy of surface molecules is higher than that of interior molecules. This excess free energy is termed the surface excess.

The directions of surface tension are different for the surfaces of different shapes. The surface tension is parallel to and falls on the surface for a flat surface, whereas the surface tension is in the tangential direction for a curved surface.

8.1.3 Surface Tensions of Common Liquids

The surface tensions of common liquids at 20°C are listed in Table 8.1.

It can be seen that the surface tension of mercury is the highest, and the surface tension of water is also quite high. Since the sphere has the least surface area in all kinds of geometry shapes, it has the least free surface energy in all kinds of shapes of liquids as well. This is why the mercury droplet and water droplet often become spherical.

8.1.4 Variation of Surface Tension With Temperature and Pressure

The surface tension of most liquids decreases with increasing temperature in a nearly linear fashion and becomes very small in the region of the critical

TABLE 8.1 Surface Tensions of Common Liquids (20°C)

Liquid	Water	Benzene	CCl_4	Acetic acid	Acetone	Ethanol	*n*-Octanol	*n*-Hexane	*n*-Octane	Mercury
γ (mN m^{-1})	72.75	28.88	26.8	27.6	23.7	22.3	27.5	18.4	21.8	485

temperature when the intermolecular cohesive forces approach zero. A number of empirical equations have been suggested which relate surface tension and temperature; one of the most satisfactory being that of Ramsay and Shields

$$\gamma\left(\frac{M}{\rho}\right)^{2/3} = K(T_c - T - 6) \tag{8.3}$$

where M is the molecular weight of the liquid, ρ the density of the liquid, T_c the critical thermodynamic temperature, T the thermodynamic temperature, K a constant, and $(M/\rho)^{2/3}$ reflects the size of the surface area per mole of liquid.

The surface tension of most liquids decreases with increasing pressure because the increase of pressure will result in the increase in the density of air, thus decreasing the net inward pull.

Example 8.1 Surface Tension and Surface Free Energy

The surface tension of mercury at 20°C $\gamma = 4.85 \times 10^{-1}$ J m^{-2}. A mercury droplet of radius (r_1) of 1 mm is divided into the droplets of radius (r_2) of 10^{-5} mm at the pressure of 101.325 kPa and the temperature of 25°C. Calculate the reversible work which will be required.

Solution

Let σ_1 be the surface area of the mercury droplet before dividing and σ_2 be the total surface area of mercury droplets after dividing. The reversible work is given by

$$W' = \int_{\sigma_1}^{\sigma_2} \gamma\, d\sigma = \gamma(\sigma_2 - \sigma_1)$$

where

$$\sigma_1 = 4\pi r_1^2; \quad \sigma_2 = N(4\pi r_2^2)$$

$$N = \left(\frac{r_1}{r_2}\right)^3$$

$$\sigma_2 = 4\pi r_2^2\left(\frac{r_1}{r_2}\right)^3 = 4\pi r_1^2\left(\frac{r_1}{r_2}\right)$$

thus

$$W' = \gamma \times 4\pi r_1^2\left(\frac{r_1}{r_2} - 1\right) = 4.85\times10^{-1}\times4\times3.1416\times\left(1\times10^{-3}\right)^2\times\left(10^5 - 1\right)$$
$$= 6.09\times10^{-1}(\text{J})$$

The surface tension and surface free energy are the most important contents of colloid and interface chemistry. It is well known that water has the highest surface tension, except for mercury, in all kinds of liquids at common temperature, and reaches up to 72.75 mN m^{-1} at 20°C. Whereas the surface tensions of other liquids are generally in the range of 20~50 mN m^{-1}. Therefore the interface characteristics of water are very significant, for example, capillary phenomenon, adsorption, and wetting, and play a very important role in water treatment. The interface characteristics of water depend on the molecular structure of water. The molecule of water has very strong polarity and very strong ability to form a hydrogen bond, which gives the water very strong cohesion force that reaches up to 22,000 atm, whereas the cohesion forces of other liquids are generally in the range of 2,000~5,000 atm, which are much smaller than that of water. The strong cohesion of water pulls the molecules at the surface into the interior of the water, producing very high surface tension.

8.2 RELATION BETWEEN LIQUID PRESSURE AND SURFACE CURVATURE

It was found that there is a balancing pressure difference across any curved liquid surface, which is discussed later. Let the pressure of gas phase be P_0, the pressure of liquid phase P', and for a plate gas–liquid interface at equilibrium

$$P_0 = P' \tag{8.4}$$

Now let us consider the case of curved gas–liquid interface, as show in Fig. 8.3. For a convex surface ABC, since the surface tension is in the tangential direction of the surface, the composite force of the surface tensions at the perimeter points to the interior of the liquid. As a result, the pressure P' must be greater than P_0 and is given by

$$P' = P_0 + \Delta P \tag{8.5}$$

where ΔP is termed extra stress and is positive. For a concave surface A′B′C′, the composite force of the surface tensions at the perimeter points to the interior of the gas phase. As a result, the pressure of gas phase P'' must be smaller than P_0, that is, the value of ΔP is negative.

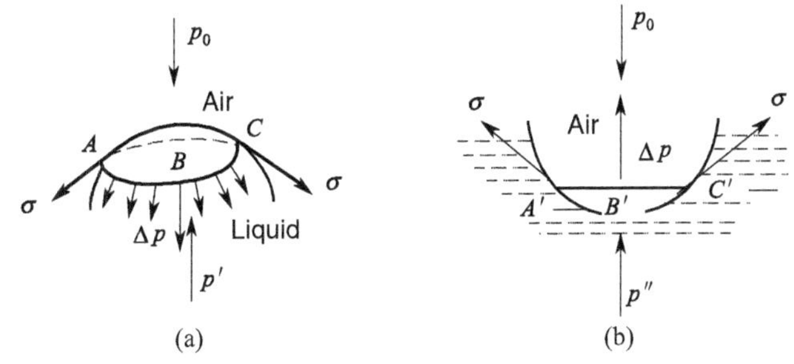

FIGURE 8.3 Extra stress for curved surface: (a) convex surface and (b) concave surface.

From the basic concept introduced earlier, the surface free energy of a spherical droplet with radius of r can be expressed as $G = 4\pi r^2\gamma$. Assuming this spherical droplet contracts so that its radius decreases by dr, then the decrease of free energy of the system is given by

$$dG = 8\pi r\gamma dr \tag{8.6}$$

The work which will be done to the environment in this process is given by

$$\delta W' = \Delta P \cdot 4\pi r^2 dr \tag{8.7}$$

The decrease in free energy of the system should equal the work done to the environment, that is,

$$\Delta P \cdot 4\pi r^2 dr = 8\pi r\gamma dr \tag{8.8}$$

thus

$$\Delta P = \frac{2\gamma}{r} \tag{8.9}$$

Eq. (8.9) is discussed as follows:

1. For convex liquid surface: since $r > 0$, $\Delta P > 0$, therefore, the pressure inside liquid is greater than that outside liquid.
2. For flat surface: since $r = \infty$, $\Delta P = 0$, therefore, the pressure inside liquid is equal to that outside liquid.
3. For concave surface: since $r < 0$, $\Delta P < 0$, therefore, the pressure inside liquid is smaller than that outside liquid.
4. For soap bubble: since there are two surfaces (inner surface and outer surface), $\Delta P = \frac{4\gamma}{r}$, which is two times Eq. (8.9).
5. For any curved surface with principal radii of curvature r_1 and r_2:

$$\Delta P = \gamma\left(\frac{1}{r_1} + \frac{1}{r_2}\right) \tag{8.10}$$

Eq. (8.10) is referred to as the Yang–Laplace equation. For a spherical surface, r_1 equals r_2, then Eq. (8.10) becomes Eq. (8.8). This equation gives an explanation for some interesting phenomena, for example, if there is a layer of water between two flat glasses, it will be hard to separate these two plate glasses, as shown in Fig. 8.4. Commonly, it can be observed that there is a concave surface at the air–water interface between two plats. Let δ be the thickness of the water layer, then the principal radii of curvature $r_1 = \delta/2$ and $r_2 = \infty$; the extra stress of the interior of water is given by

$$\Delta P = \gamma\left(\frac{1}{-\delta/2} + \frac{1}{\infty}\right) = -\frac{2\gamma}{\delta} < 0$$

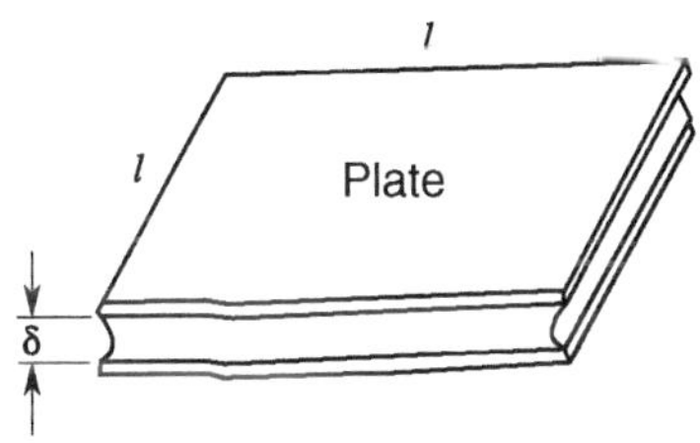

FIGURE 8.4 **Capillary pressure.**

It implies the pressure of the water layer is smaller than that of air, and their pressure difference makes the separation described earlier difficult. This pressure difference is referred to as capillary pressure.

Example 8.2 Extra Stress of Curved Surface

There are two soap bubbles of different sizes which are connected to each other by a tube with a valve, as shown in the next figure. If the valve is opened, what will happen? And how will the two bubbles change?

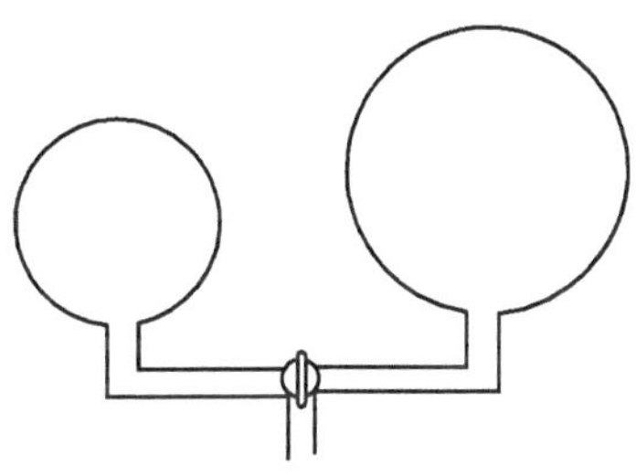

Solution

Since the curvature radius of the small bubble on the left is smaller than that of the big bubble on the right, according to the Yang–Laplace equation, the extra stress of the small bubble is greater than that of the big bubble, thus the gas in the small bubble will flow to the big bubble, and the small bubble will become smaller and smaller with quickly increasing curvature radius. Meanwhile, the big bubble will become bigger and bigger with slowly increasing curvature radius until the curvature radii gradually become equal and the system reaches equilibrium.

Example 8.3 Extra Stress of Curved Surface

A capillary tube with a diameter of 0.1 cm is inserted 10 cm deep into a dilute solution to blowout an air bubble at the end of the tube. The maximum pressure ΔP required equals to that of 11.6 cm high water column. Calculate the surface tension of this solution.

Solution

Suppose the density of the solution is the same as that of pure water. According to the Yang–Laplace equation

$$\Delta P = \frac{2\gamma}{r}$$

$$(11.6-10.0)\times 0.01\times\frac{101325}{10.336}=\frac{2\gamma}{r}$$

$$156.85=\frac{2\gamma}{0.0005}$$

$$\gamma = 0.0392\ (\mathrm{N\,m^{-1}}) = 39.2\ (\mathrm{mN\,m^{-1}})$$

8.3 RELATION BETWEEN VAPOR PRESSURE OF LIQUID AND SURFACE CURVATURE

The surface area of a spherical droplet is known as $s = 4\pi r^2$, where r is the radius of the droplet. If the radius of a droplet increases from r to $r + dr$, the increase in surface area will be

$$ds = 8\pi r dr \tag{8.11}$$

thus the increase in surface free energy will be given by

$$dG = 8\pi r\gamma dr \tag{8.12}$$

If this process involves the transfer of dn moles of liquid from the planar surface with a vapor pressure P_0 to the droplet with a vapor pressure P_r, as shown in Fig. 8.5, the chemical potential of liquid before the transfer will be

$$\mu = \mu^{\ominus} + RT\cdot\ln\frac{P_0}{P^{\ominus}} \tag{8.13}$$

and the chemical potential of liquid after the transfer will be

$$\mu' = \mu^{\ominus} + RT\cdot\ln\frac{P_r}{P^{\ominus}} \tag{8.14}$$

where μ is the chemical potential of the plane liquid and μ' the chemical potential of the droplet. Thus the change in free energy is given by

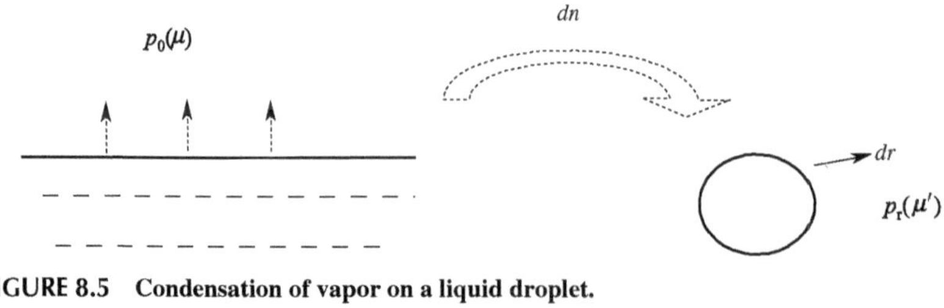

FIGURE 8.5 **Condensation of vapor on a liquid droplet.**

$$dG = dn(\mu' - \mu) = dnRT \cdot \ln\frac{P_r}{P_0} \tag{8.15}$$

Equating Eqs. (8.15) and (8.12) gives

$$dnRT \cdot \ln\frac{P_r}{p_0} = 8\pi r\gamma dr \tag{8.16}$$

Let ρ be the density of the liquid and M the molecular weight of the liquid

$$dn = 4\pi r^2 dr\left(\frac{\rho}{M}\right) \tag{8.17}$$

Substituting Eq. (8.17) for dn in Eq. (8.16) gives

$$RT \cdot \ln\frac{P_r}{p_0} = \frac{2\gamma M}{r\rho} = \frac{2\gamma V}{r} \tag{8.18}$$

where V is the molar volume of the liquid and r is termed the radius of curvature. This expression, known as the Kelvin equation, has been verified experimentally.

From the Kelvin equation, we know the following:

1. The smaller the radius of droplet, the greater the vapor pressure; therefore, water vapor can reach a very high degree of supersaturation in clear air.
2. For a convex surface: Since $r > 0$, $P_r > P_0$.
3. For a concave surface: Since $r < 0$, $P_r < P_0$.

Example 8.4 Relation Between Vapor Pressure of Liquid and Surface Curvature

If water vapor is sharply cooled down to 25°C, the supersaturation of water vapor will take place. The surface tension of water at 25°C is known as 71.49 mN m^{-1}. Suppose the pressure of supersaturated vapor is four times that of water at 25°C, calculate

1. the radius of initial water droplet under this supersaturation;
2. the number of molecules contained in one initial droplet.

Solution

According to the Kelvin equation

1. $\ln\frac{P_r}{P_0} = \frac{M}{\rho} \cdot \frac{2\gamma}{RTr}$

   $$\ln 4 = \frac{0.018}{1000} \cdot \frac{2 \times 71.49 \times 10^{-3}}{8.314 \times 298r}$$

   $r = 0.75 \times 10^{-9}$ (m) $= 0.75$ (nm)

2. $n = \frac{4}{3}\pi r^3 \rho \times \frac{1}{M} \times N_A = \frac{4}{3} \times 3.14 \times (0.75 \times 10^{-9})^3 \times 1000 \times \frac{1}{0.018} \times 6.02 \times 10^{23}$
   $= 59$

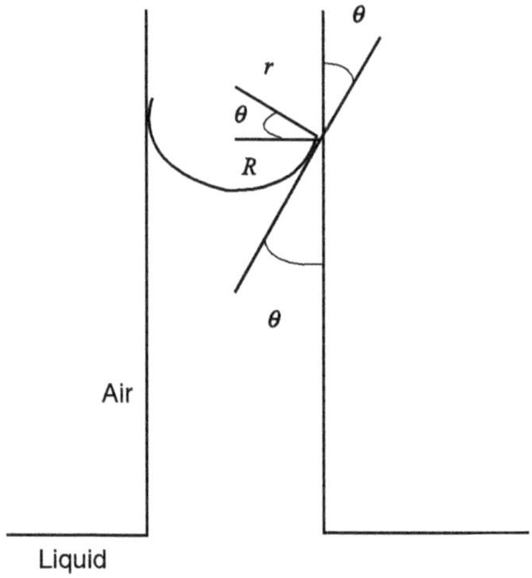

FIGURE 8.6 **Capillary rise.**

8.4 CONTACT ANGLE

A quantity that is closely related to surface tension is the contact angle. The contact angle θ is defined as the angle (measured in the liquid) that is formed at the junction of three phases, for example, at the solid–liquid–gas junction as shown in Fig. 8.6. Although the surface tension is the property of the two phases that form the interface, θ requires that three phases be specified for its characterization. The definition of contact angle is, however, highly simplified, and we take a more in-depth look at the concept later in chapter: Surface of Solids.

8.5 MEASUREMENT OF SURFACE TENSION

There are many methods for measuring surface tension, for example, capillary rise method, Wilhelmy plate method, ring method, drop volume and drop weight method, pendant drop method, and oscillating jet method. Here we introduce only the capillary rise method and the ring method, which are used commonly.

8.5.1 Capillary Rise Method

Capillary rise method, when properly performed, is the most accurate method available for determining surface tensions. This method is based on the Yang–Laplace equation described previously.

First let us insert a glass capillary tube into a liquid, as shown in Fig. 8.6. Since a curved meniscus forms in the capillary tube, the extra pressure ΔP is smaller than zero, that is, the liquid pressure inside the tube is smaller than that outside the tube, and the difference is ΔP, thus leading to the rise of liquid up this narrow capillary until the static pressure of liquid column equals the pressure difference. Let ρ be the density of the liquid, ρ' the density of the air, h the height of the liquid column, and g the acceleration of gravity, then we have

$$\Delta P = \rho gh - \rho' gh = (\rho - \rho')gh$$

Combining this equation with the Yang–Laplace equation and neglecting the sign of ΔP gives

$$\frac{2\gamma}{r} = (\rho - \rho')gh$$

or

$$\gamma = \frac{r(\rho - \rho')gh}{2} \tag{8.19}$$

where r is the radius of curvature, but it is more convenient to calculate γ in terms of the radius of capillary tube. For this purpose, let us make a tangent line of the air–liquid interface through the junction of three phases, thus we will obtain the contact angle θ. After that, we draw the radius of curvature r and the radius of capillary tube R in the figure. It can be proven that the angle between r and R equals the contact angle θ. Thus,

$$r = \frac{R}{\cos\theta} \tag{8.20}$$

Substituting Eq. (8.20) for r in Eq. (8.19) gives

$$\gamma = \frac{R(\rho - \rho')gh}{2\cos\theta} \tag{8.21}$$

Eq. (8.21) is the basis of the capillary rise method for measuring the contact angle. Suppose $\theta = 0°$ and $\rho \gg \rho'$, the approximate equation is obtained as follows

$$\gamma = \frac{Rh\rho g}{2} \tag{8.22}$$

Using Eq. (8.22), the surface tension can be conveniently obtained by measuring only the height of the liquid column at equilibrium.

In addition, we have another simple way to derive the expression for the capillary rise method. At equilibrium the vertical component of the surface tension must be equal to the weight of the liquid column, approximated as the weight of a cylinder of height h and radius R. This leads to the approximation

$$2\pi R\gamma\cos\theta = \pi R^2 h(\rho - \rho')g \tag{8.23}$$

Here $(\rho - \rho')$ is the difference in the density of the liquid and its surroundings and is used in the formulation, because the surrounding fluid can be a second liquid that has a buoyant effect on the liquid column. Rearranging Eq. (8.23), we obtain

$$\gamma = \frac{R(\rho - \rho')gh}{2\cos\theta} \approx \frac{R\rho gh}{2}$$

8.5.2 Ring Method

In this method the force required to detach a ring from a surface or interface is measured either by suspending the ring from the arm of a balance or by using a torsion wire arrangement, as in Fig. 8.7.

The detachment force is related to the surface or interfacial tension by the expression

$$\gamma = \frac{\beta F}{4\pi R} \tag{8.24}$$

where F is the pull on the ring, R the mean radius of the ring, and β a correction factor.

To ensure zero, and hence a constant contact angle, platinum rings carefully cleaned in strong acid or by flaming are used. It is essential that the ring should lie flat in a quiescent surface. For interfacial work, the lower liquid must wet the ring preferentially, for example, for benzene on water, a clean platinum ring is suitable, whereas for water on carbon tetrachloride, the ring must be made hydrophobic.

The correction factor β allows for the nonvertical direction of the tension force and for the complex shape of the liquid supported by the ring at the point of detachment; hence it depends on the dimensions of the ring and the nature of the interface. The values can be obtained by consulting the specialized table in other works.

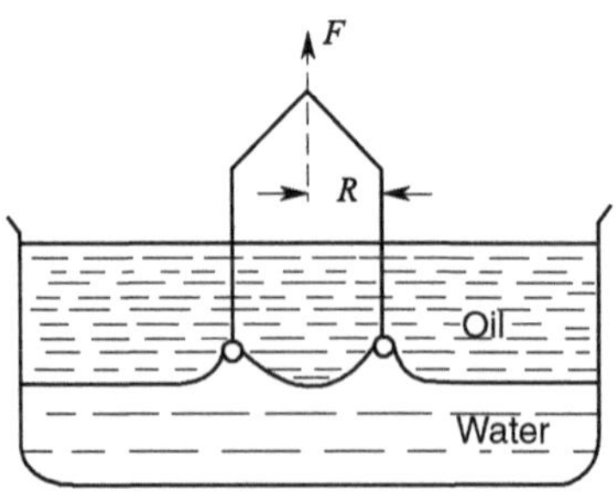

FIGURE 8.7 Measurement of interfacial tension by the ring method.

Example 8.5 Measurement of Interfacial Tension

The oil–water interfacial tension is measured by the capillary rise method, as shown in the following figure. The radius of the capillary tube is 0.1 mm, the height of water column at equilibrium 4 cm, the density of oil 0.8 g cm^{-1}, and the contact angle of glass–water–oil intersection 40 degree. Calculate the oil–water interfacial tension.

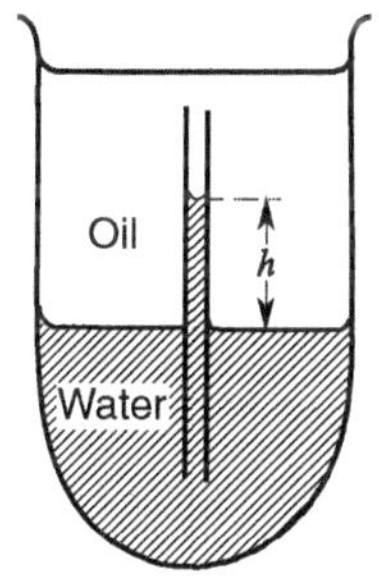

Solution

According to Eq. (8.21)

$$\gamma = \frac{R(\rho - \rho')gh}{2\cos\theta} = \frac{0.1\times10^{-3}(1000-800)\times9.8\times4\times10^{-2}}{2\times0.766}$$
$$= 5.1\times10^{-3}\ (\text{N m}^{-1}) = 5.1\,(\text{mN m}^{-1})$$

8.6 COHESION WORK AND ADHESION WORK

Here we discuss cohesion work and adhesion work of liquids. Assuming there are two separation processes for liquids, the schematic illustrations of these two processes are shown in Fig. 8.8. In Fig. 8.8a, which applies to a pure liquid, the process consists of producing two new interfaces, each of a unit cross-section. Therefore, for the separation process,

$$W_{AA} = 2\gamma_A \tag{8.25}$$

where the quantity W_{AA} is known as the work of cohesion because it equals the work required to pull apart a column of liquid A. It represents the increase in surface free energy of the system in this process and measures the attraction between the molecules of the two portions.

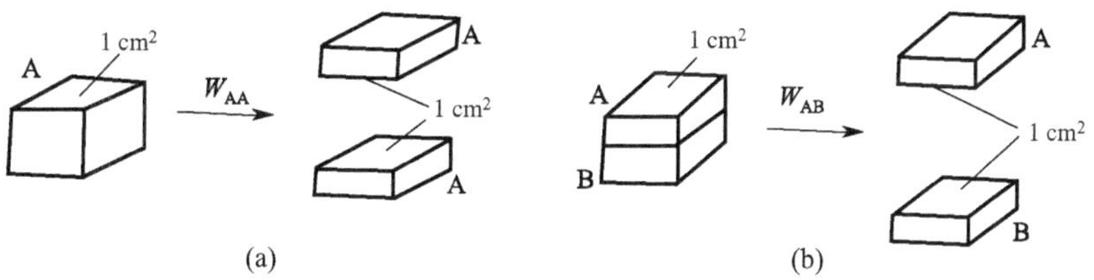

FIGURE 8.8 (a) Cohesion work and (b) adhesion work.

Now let us consider the separation of A and B as represented in Fig. 8.8b, each of a unit cross-section. For this process,

$$W_{AB} = \gamma_A + \gamma_B - \gamma_{AB} \tag{8.26}$$

where the quantity W_{AB} is known as the work of adhesion. It equals the work required to separate the liquid A and liquid B and form two separate liquid–air interfaces. It represents the increase in surface free energy of the system in this process and measures the attraction between the two different phases.

8.7 SPREADING OF ONE LIQUID ON ANOTHER LIQUID

When a drop of oil is placed on a clean water surface it may behave in one of three ways:

1. remaining as a lens, as in Fig. 8.9a;
2. spreading as a thin film, which may show interference colors, until it is uniformly distributed over the surface as a duplex film (a duplex film is a film which is thick enough for the two interfaces, ie, liquid–film and film–air, to be independent and possess characteristic surface tension);
3. spreading as a monomolecular layer, leaving excess oil as lens in equilibrium, as in Fig. 8.9b.

If the area occupied by the oil drop is increased by dA, the change in the surface free energy of the system will be approximately

$$dG = (\gamma_{O/W} + \gamma_{O/A} - \gamma_{W/A})dA \tag{8.27}$$

If this quantity is negative, the process of spreading will take place spontaneously. Generally, the term initial spreading coefficient (for the case of A on B) is defined as

$$S_{A/B} = \gamma_B - \gamma_A - \gamma_{A/B} \tag{8.28}$$

thus we have criterion as

$$S_{A/B} > 0 \text{ A spreads on B spontaneously.} \tag{8.29}$$

$$S_{A/B} < 0 \text{ A does not spread on B spontaneously.} \tag{8.30}$$

Generally speaking, it is easy for the liquid of lower surface tension to spread on the liquid of higher surface tension.

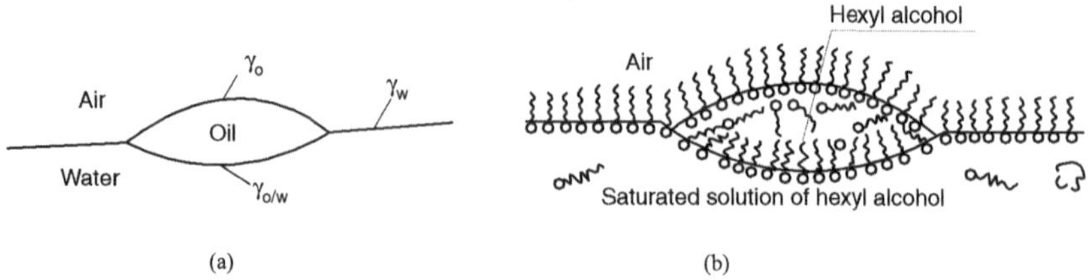

FIGURE 8.9 Spreading of one liquid on another liquid: (a) formation of lens and (b) formation of monomolecule film.

Example 8.6 Spreading of One Liquid on Another

We know that $\gamma_{water} = 73$ mN m^{-1}, $\gamma_{mercury} = 485$ mN m^{-1}, $\gamma_{octanol} = 27$ mN m^{-1}, $\gamma_{water/mercury} = 375$ mN m^{-1}, $\gamma_{water/octanol} = 9$ mN m^{-1}, and $\gamma_{mercury/octanol} = 348$ mN m^{-1} at common temperatures. Calculate the following: (1) initial spreading coefficient when water is placed on mercury surface, octanol surface, and mercury/octanol surface, respectively; and (2) the initial spreading coefficient when octanol is placed on water surface, mercury surface, and mercury/water surface, respectively.

Solution

According to Eq. (8.28)

1. $S_{water/mercury} = \gamma_{mercury} - \gamma_{water} - \gamma_{water/mercury} = 485 - 73 - 375 = 37 \ (\text{mN m}^{-1})$
   $S_{water/octanol} = \gamma_{octanol} - \gamma_{water} - \gamma_{water/octanol} = 27 - 73 - 9 = -55 \ (\text{mN m}^{-1})$
   $S_{water/(mercury/octanol)} = \gamma_{mercury/octanol} - \gamma_{water/mercury} - \gamma_{water/octanol} = 348 - 375 - 9 = -36 \ (\text{mN m}^{-1})$

2. $S_{octanol/water} = \gamma_{water} - \gamma_{octanol} - \gamma_{octanol/water} = 73 - 27 - 9 = 37 \ (\text{mN m}^{-1})$
   $S_{octanol/mercury} = \gamma_{mercury} - \gamma_{octanol} - \gamma_{octanol/mercury} = 485 - 27 - 348 = 110 \ (\text{mN m}^{-1})$
   $S_{octanol/(mercury/water)} = \gamma_{mercury/water} - \gamma_{octanol/water} - \gamma_{octanol/mercury} = 375 - 9 - 348 = 18 \ (\text{mN m}^{-1})$

Example 8.7 Spreading of One Liquid on Another

The interfacial tensions of interfaces formed by water and isopentanol before dissolution of each other and after dissolution of each other are shown respectively in the next table.

Interface	Water/air	Isopentanol/air	Water/isopentanol
γ (mN m^{-1}) before dissolution	72.8	23.7	5.0
γ (mN m^{-1}) after dissolution	25.9	23.6	5.0

Place a drop of isopentanol on the water surface. What will happen at the beginning and after some time? Give the explanation for these phenomena.

Solution

At the beginning

$$S = \gamma_{water/air} - \gamma_{water/isopentanol} - \gamma_{isopentanol/air} = 72.8 - 5.0 - 23.7 = 44.1 \ (\text{mN m}^{-1}) > 0$$

thus the spreading will take place spontaneously.

After some time, the interfacial tensions are reduced due to the dissolution of each other, thus

$$S = 25.9 - 23.6 - 5.0 = -2.7 \ (\text{mN m}^{-1}) < 0$$

Therefore, the isopentanol will aggregate on the water surface.

8.8 FOWKES THEORY OF INTERFACIAL TENSION

Throughout this chapter we have dealt with surface tension from a macroscopic point of view almost exclusively. From a fundamental perspective, however, descriptions from a molecular perspective are very difficult. For liquid–liquid interfacial tension, Fowkes et al. developed a semiempirical treatment, which is valuable to introduce here.

As described previously, surface tension is caused by a net inward pull, that is, a vertically inward composite force which involves intermolecular forces, including hydrogen bonding (h), metallic bonding (m), dispersion force (d), electro interaction (π), and ion interaction (i); therefore, surface tension is given by

$$\gamma = \gamma^{\mathrm{d}} + \gamma^{\mathrm{h}} + \gamma^{\mathrm{m}} + \gamma^{\pi} + \gamma^{\mathrm{i}} \tag{8.31}$$

in which only the dispersion component of intermolecular attraction is ubiquitous and the most long distance one compared with other interactions, thus, operates across the interface to decrease the work required to bring a particle to the surface. This decrease in the work ($-\Delta G$) equals the geometric mean of dispersion components of liquid A and liquid B which form the interface.

$$-\Delta G = (\gamma_{\mathrm{A}}^{\mathrm{d}} \gamma_{\mathrm{B}}^{\mathrm{d}})^{1/2} \tag{8.32}$$

Accordingly the work required to bring molecule A to interface AB is given by $\gamma_{\mathrm{A}} - (\gamma_{\mathrm{A}}^{\mathrm{d}} \gamma_{\mathrm{B}}^{\mathrm{d}})^{1/2}$; similarly, the work required to bring molecule B to interface AB is given by $\gamma_{\mathrm{B}} - (\gamma_{\mathrm{A}}^{\mathrm{d}} \gamma_{\mathrm{B}}^{\mathrm{d}})^{1/2}$, thus the total work required to form the interface AB of unit area is the sum of these contributions

$$\gamma_{\mathrm{AB}} = \gamma_{\mathrm{A}} + \gamma_{\mathrm{B}} - 2\,(\gamma_{\mathrm{A}}^{\mathrm{d}} \gamma_{\mathrm{B}}^{\mathrm{d}})^{1/2} \tag{8.33}$$

Combining Eqs. (8.33) and (8.26) gives

$$W_{\mathrm{AB}} = 2(\gamma_{\mathrm{A}}^{\mathrm{d}} \gamma_{\mathrm{B}}^{\mathrm{d}})^{1/2} \tag{8.34}$$

which shows

1. The work of adhesion measures the strength of attraction between two different phases indeed as described before.
2. The surface tension which operates across the interface is the dispersion component of it, that is, γ^{d}.

After careful selection and evaluation, for hydrocarbon (H), water (W), and mercury (Hg) Fowkes proposed

$$\gamma_{\mathrm{H}} = \gamma_{\mathrm{H}}^{\mathrm{d}} \tag{8.35}$$

$$\gamma_{\mathrm{W}} = \gamma_{\mathrm{W}}^{\mathrm{d}} + \gamma_{\mathrm{W}}^{\mathrm{h}} \tag{8.36}$$

TABLE 8.2 Experimental Values for γ_H, $\gamma_{W/H}$, and $\gamma_{Hg/H}$, and Calculated Values for γ_W^d and γ_{Hg}^d (20°C)

Hydrocarbon	γ_H (mJ m^{-2})	$\gamma_{Hg/H}$ (mJ m^{-2})	γ_{Hg}^d (mJ m^{-2})	$\gamma_{W/H}$ (mJ m^{-2})	γ_W^d (mJ m^{-2})
n-Hexane	18.4	378	210	51.1	21.8
n-Heptane	18.4	—	—	50.2	22.6
n-Octane	21.8	375	199	50.8	22.0
n-Nonane	22.8	372	199	—	—
n-Decane	23.9	—	—	51.2	21.6
n-Tetradecane	25.6	—	—	52.2	20.8
Cyclohexane	25.5	—	—	50.2	22.7
Decalin	29.9	—	—	51.4	22.0
Benzene	28.85	363	194	—	—
Toluene	28.5	359	208	—	—
o-Xylene	30.1	359	200	—	—
m-Xylene	28.9	357	211	—	—
p-Xylene	28.4	361	203	—	—
n-Propylbenzene	29.0	363	194	—	—
n-Butylbenzene	29.2	363	193	—	—
Average		200 ± 7		21.8 ± 0.7	

$\gamma_{Hg} = 484$ mJ m^{-2}; $\gamma_W = 72.8$ mJ m^{-2}.

$$\gamma_{Hg} = \gamma_{Hg}^d + \gamma_{Hg}^m \quad (8.37)$$

Therefore, for mercury–hydrocarbon interface, Eq. (8.33) becomes

$$\gamma_{Hg/H} = \gamma_{Hg} + \gamma_H - 2(\gamma_{Hg}^d \gamma_H^d)^{1/2} \quad (8.38)$$

in which all of interfacial tensions can be experimentally measured except for γ_{Hg}^d, thus it provides us the method of measuring γ_{Hg}^d. Similarly, for water–hydrocarbon interface we have

$$\gamma_{W/H} = \gamma_W + \gamma_H - 2(\gamma_W^d \gamma_H^d)^{1/2} \quad (8.39)$$

It provides us the method of measuring γ_W^d.

Table 8.2 lists the values of γ_H, $\gamma_{W/H}$, and $\gamma_{Hg/H}$ obtained from experiments and the calculated values of γ_W^d and γ_{Hg}^d for a number of hydrocarbons forming interfaces with water and mercury.

It shows that the average values of γ^d for water and mercury are 21.8 ± 0.7 and 200 ± 7, respectively. Based on these two values we can work out $\gamma_{Hg/w}$ by using Eq. (8.33). The result is 426 mJ m^{-2}, which is in accordance with the experimental value, thus indicates the success of the average values of γ^d in their ability to calculate γ correctly for the water–mercury interface.

Example 8.8 Fowkes Theory of Interfacial Tension

At 20°C, for a compound Y and water (W), the interface tension is known as 421.5 mN m^{-1}; for the same compound Y and a hydrocarbon (H), the interface tension is known as 378.1 mN m^{-1}; and for the same hydrocarbon and water, the interface tension is known as 50.1 mN m^{-1}. In addition, the surface tensions of water and the compound Y are 72.8 and 484.1 mN m^{-1}, respectively. Evaluate the surface tension of the hydrocarbon at 20°C.

Solution

According to Fowkes theory

$$\gamma_{WH} = \gamma_W + \gamma_H - 2\sqrt{\gamma_W^d \gamma_H^d}$$

$$\gamma_{YH} = \gamma_Y + \gamma_H - 2\sqrt{\gamma_Y^d \gamma_H^d}$$

$$\gamma_{YW} = \gamma_Y + \gamma_W - 2\sqrt{\gamma_Y^d \gamma_W^d}$$

that is,

$$51.1 = 72.8 + \gamma_H - 2\sqrt{\gamma_W^d \gamma_H^d} \quad \text{(i)}$$

$$378.1 = 484 + \gamma_H - 2\sqrt{\gamma_Y^d \gamma_H^d} \quad \text{(ii)}$$

$$421.5 = 484 + 72.8 - 2\sqrt{\gamma_Y^d \gamma_W^d} \quad \text{(iii)}$$

Considering $\gamma_H = \gamma_H^d$ and combining Eqs. (ii) and (iii) gives

$$\frac{484 - 375.1 + \gamma_H}{2\sqrt{\gamma_H}} = \sqrt{\gamma_Y^d} = \frac{484 - 421.5 + 72.8}{2\sqrt{\gamma_W^d}}$$

that is,

$$\sqrt{\gamma_W^d} = \frac{135.5\sqrt{\gamma_H}}{105.9 + \gamma_H} \quad \text{(iv)}$$

Substituting Eq. (iv) into Eq. (i) gives

$$\gamma_H^2 - 143\gamma_H + 2298 = 0$$

Solving it gives

$$\gamma_H = 18.5 \text{ mN m}^{-1}$$

and

$$\gamma_H = 124.5 \text{ mN m}^{-1} \text{(not reasonable)}$$

therefore

$$\gamma_H = 18.5 \text{ mN m}^{-1}$$

8.9 INSOLUBLE MONOMOLECULAR FILM

Many insoluble substances, such as long-chain fatty acids and alcohols, can (with the aid of suitable solvent) be spread onto a water surface, and if space permits, will form a surface film one molecule in thickness with the hydrophilic $-COOH$ or $-OH$ groups oriented toward the water phase, and the hydrophobic hydrocarbon chains oriented away from the water phase. This kind of film is referred to as insoluble monomolecular film or monolayer.

8.9.1 Surface Pressure

It has been well known since ancient times that oil has the ability to overcome waves on the sea, therefore, lifeboats often carried fish oil with them. The reason is that insoluble monomolecular film has surface pressure. The surface pressure is observed in many phenomena, for example, make a loop with a thin wire and place it on the surface of water, then place a droplet of oleic acid in the center of the loop. You will find that the loose loop becomes tight immediately, as shown in Fig. 8.10a. Place a match on the surface of the water, then, with the end of another match that had been scraped against a person's face, dip it into the surface of water beside the first match. You will find that the match on the surface moves toward the other side, as shown in Fig. 8.10b. In both of the experiments the monolayer forms. It applies a force to the floating material, thus leading to the phenomena described previously. Commonly, the force applied to the unit length of floating material is termed surface pressure and generally represented by π.

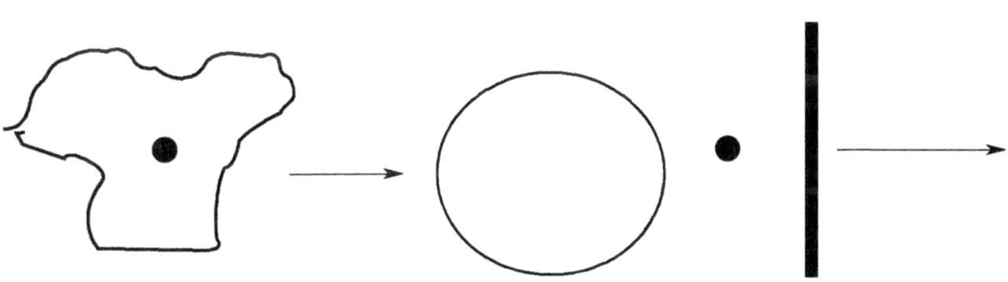

(a) (b)

FIGURE 8.10 **Instance of surface pressure: (a) loose loop becomes tight and (b) match on the surface moves.**

Let the length of the match be L, the moving distance be dx, the surface tension of clean water be γ_0, and the surface tension of the monolayer be γ. Since the decrease in surface free energy equals the work required to move the match, we have

$$\pi L dx = (\gamma_0 - \gamma) L dx \tag{8.40}$$

thus

$$\pi = \gamma_0 - \gamma \tag{8.41}$$

Eq. (8.41) indicates that the surface pressure equals the decrease in surface tension due to the increasing unit area of the film.

Example 8.9 Surface Pressure

Suppose the surface pressure of a monolayer film π is 50 mN m^{-1}, the length of a floating material is 1 m, and the thickness of the film is 1 nm, calculate the force applied to the unit area of cross-section of the floating material.

Solution

The force applied to the unit area of cross-section of the floating material is given by

$$F = \frac{50}{1\times10^{-9}} = 5\times10^{10}\ (\text{mN m}^{-2}) \approx 500\ \text{atm}$$

which shows the surface pressure is quite high.

8.9.2 Various States of Monomolecular Film

Two-dimensional monomolecular film or monolayers can exist in different physical states which bear some resemblance to the solid, liquid, and gaseous states in three-dimensional matter. These surface films are best classified according to the lateral adhesion between the film molecules, including hydrocarbon chains and end-groups. Here we discuss the various states in the sequence they appear when the film is compressed, as shown in Fig. 8.11.

1. Gaseous films

   The line G in Fig. 8.11 represents gaseous film in which the molecules occupy larger areas and have smaller surface pressure. For ideal states, the state equation is given by

   $$\pi a = K_B T \tag{8.42}$$

   where a is the area occupied by a molecule, K_B the Boltzmann constant, and T the thermodynamic temperature. This equation is very similar in form to

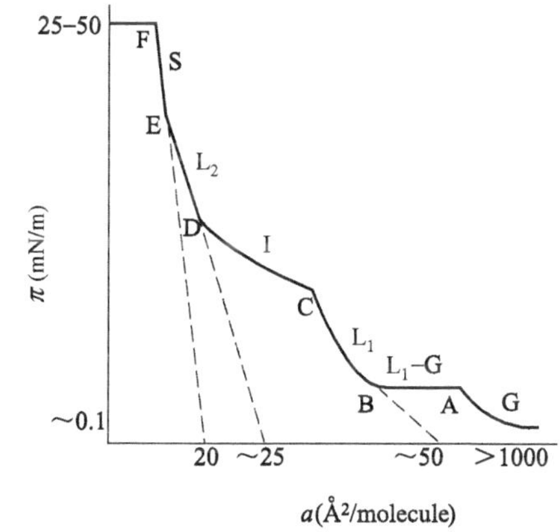

FIGURE 8.11 **Various states of insoluble monomolecular film.**

the P–V relation for ideal gas and is referred to as the ideal two-dimensional gas equation. For the actual film, the expression is given by

$$\pi(a - a_0) = iK_B T \tag{8.43}$$

or

$$\left(\pi + \frac{b}{a^2}\right)(a - a_0) = K_B T \tag{8.44}$$

where a_0 is the excluded area per molecule and both i and b are constants.

2. Gas–liquid coexistence region
If the gaseous film is gradually compressed, the area occupied by a molecule will be gradually decreased. When the line reaches point A, the surface pressure will not rise until reaching point B. The horizontal line of L_1-G from A to B is referred to as gas–liquid coexistence region or gas–liquid equilibrium line. This line is analogous in every way to the corresponding feature in bulk matter. The film pressures at which gas–liquid equilibrium occurs are known as film vapor pressure.
3. Liquid-expanded film
As the film is continuously compressed, the gas phase begins condensing to a liquid-like phase known as the liquid-expanded film, which is represented by line L_1 from B to C. Liquid-expanded film is the first of several condensed states and has greater expansion coefficient. It obeys the following equation of state:

$$\pi_p(a - a_0) = K_B T \tag{8.45}$$

where π_p is the contribution of the polar group to π and a_0 the area occupied by the polar group. Suppose the attraction between hydrocarbon chains is π_0, which is opposite to π_p in direction, then the surface pressure is the algebraic sum of two contributions as follows:

$$\pi = \pi_p + \pi_0 \tag{8.46}$$

Therefore the equation of state for liquid-expanded film is given by

$$(\pi - \pi_0)(a - a_0) = K_B T \tag{8.47}$$

4. Transition state film
   With sufficient compression the isotherm of liquid-expanded film shows a sharp break to enter a situation variously known as the intermediate or transition state, indicated by I. As its name implies, state I was thought of as an intermediate transition state between the liquid-expanded film and the liquid-condensed film state (described in the next paragraph). This film has much greater expansion coefficient, as shown in Fig. 8.11.
5. Liquid-condensed film
   If the area of an insoluble monolayer is isothermally reduced still further, the compressibility eventually becomes very low. Because of the low compressibility, the states observed at these low values of a_0 are called liquid-condensed state, which is indicated by L_2 from D to E. In this region, the isotherm is essentially linear, and the film has a small expansion coefficient, as shown in Fig. 8.11, just like common liquid.
6. Solid-condensed film
   When compressed further, the water in the liquid-condensed film will be squeezed out, forming solid-condensed film represented by line S from E to F. The isotherm is linear, and the film has a very small expansion coefficient, as shown in Fig. 8.11. Extending line L_2 and line S to $\pi = 0$ respectively shows that the section areas of molecules of the homolog with linear chain equal $25 Å^2$ per molecule and $20 Å^2$ per molecule respectively. This implies that surfactant molecule is vertically orientated at the surface in both of the states.

8.9.3 Application of Surface Film

The research of surface film not only proved that the hydrocarbon chain at the surface is oriented in a certain direction, but also provided a very efficient method for some important investigations.

1. Measurement of molecular weight of polymer
   Suppose a polymer forms a gaseous monolayer and i equals 1; its equation of state will be

$$\pi(a - a_0) = K_B T \tag{8.48}$$

Multiplying both sides of Eq. (8.48) by the mole number n and the Avogadro constant N_A gives

$$\pi(A - nA_0) = nRT = \frac{W}{M}RT \tag{8.49}$$

or

$$\pi A = nA_0\pi + \frac{W}{M}RT \tag{8.50}$$

where A is the area of the film, A_0 the area of material itself, and W the weight of the material. Plotting the straight line of πA versus π will give the molecular weight of the polymer.

2. Water conservation
The annual loss of water from hot country lakes and reservoirs due to evaporation is usually about 3 m per year. This evaporation can be considerably reduced by coating the water surface with an insoluble monolayer; for example, a monolayer of cetyl alcohol can reduce the rate of evaporation by as much as 40%. To see how this is quantified, consider Example 8.10.

Example 8.10 Monolayer

There is a reservoir with an area of 100 km^2. Suppose we want to spread a monolayer of cetyl alcohol on the surface in order to prevent the evaporation of water, and this film is liquid-condensed film. Calculate how many kilograms of cetyl alcohol will be needed.

Solution

As described previously, the area occupied by one molecule of liquid-condensed film is

$$a = 25\ \text{Å}^2 \text{ per molecule}$$

Thus, the area occupied by one mole of liquid-condensed film is

$$A_0 = 25 \times (10^{-10}\ \text{m})^2 \times 6.02 \times 10^{23} = 150.5 \times 10^3\ (\text{m}^2\ \text{mol}^{-1})$$

For cetyl alcohol

$$M = 0.242\ (\text{kg mol}^{-1})$$

Therefore, the weight of cetyl alcohol needed is given by

$$W = \frac{10^8}{150.5 \times 10^3} \times 0.242 = 160.8\ (\text{kg})$$

Chapter 9

Surface of Solution

Chapter Outline

9.1 SURFACE ACTIVITY

Experiment shows the surface tension of solution is affected by the type of solution and changes with the solute concentration. There are three types of variations of γ with c as shown in Fig. 9.1.

Curve A is typical of inorganic electrolytes and highly hydrated organic compounds. Curve B is the type of behavior characteristic of most unionized organic compounds. The type of behavior indicated by curve C is shown by soluble amphipathic species, especially ionic ones. The solute for curve C can sharply lower the surface tension of solution with smaller dosage of solute, that is, it is of surface activity. This kind of substance is referred to as a surfactant. Its surface activity is shown in terms of Eq. (9.1)

$$\frac{\gamma_0 - \gamma}{c} \tag{9.1}$$

where γ_0 is the surface tension of pure water, γ the surface tension of the solution of surfactant, and c the concentration of surfactant.

Colloid and Interface Chemistry for Water Quality Control. http://dx.doi.org/10.1016/B978-0-12-809315-3.00009-8

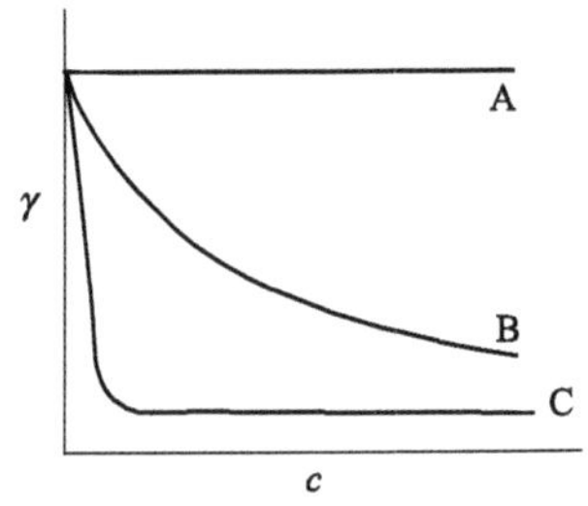

FIGURE 9.1 **Three types of variation of γ with c for aqueous solution.**

1. Traube rule
   It was observed from experiment that the longer the molecular chain, the stronger the surface activity of the surfactant. The Traube rule showed that the increase of one —CH_2— leads to three times the increase in surface activity of the surfactant.
2. Szyszkowski empirical equation
   For dilute solution of surfactant, the Szyszkowski empirical equation was proposed as follows

$$\frac{\gamma_0 - \gamma}{\gamma_0} = b \cdot \ln\left(\frac{c}{a} + 1\right) \tag{9.2}$$

where a and b are empirical constant, and b is of the same value for homologs.

Since $\ln\left(\frac{c}{a}+1\right) \approx \frac{c}{a}$

Eq. (9.2) becomes

$$\frac{\gamma_0 - \gamma}{\gamma_0} = \frac{b}{a}c \tag{9.3}$$

If $c = 1$, the b/a will represent the fraction of decrease in surface tension. Therefore, the surface activity of surfactant can be measured by b/a.

9.2 SURFACE EXCESS AND GIBBS ADSORPTION EQUATION

9.2.1 Surface Excess

Assuming there is a beaker of solution, as shown in Fig. 9.2, let the liquid phase be α, gas phase β, gas–liquid interface s–s, and a–a to b–b interfacial region be represented by σ.

Again let the mole number of solute be n, the concentration of solute (mole per volume) in gas phase or liquid phase be c, and the volume of gas phase or

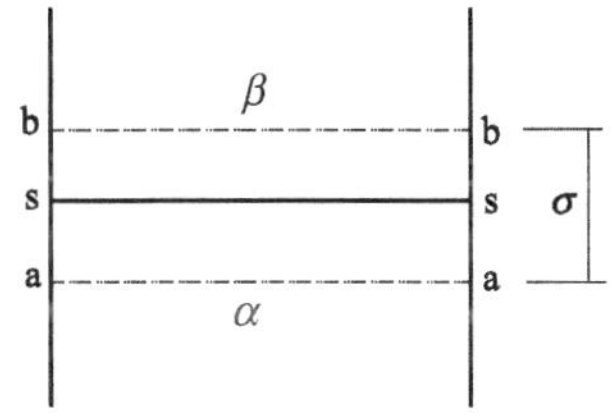

FIGURE 9.2 Interfacial region of solution.

liquid phase be V. If the concentration of solute in the interfacial region were the same as that in the bulk solutions, we would have

$$n = n^{\alpha} + n^{\beta} = c_{\alpha}V_{\alpha} + c_{\beta}V_{\beta}$$

However, they are not the same in practice because of the existence of surface excess. Therefore, we suppose

$$n^{\sigma} = n - (n^{\alpha} + n^{\beta}) = n - (c_{\alpha}V_{\alpha} + c_{\beta}V_{\beta}) \tag{9.4}$$

and

$$\Gamma = \frac{n^{\sigma}}{A} \tag{9.5}$$

Eq. (9.5) is the expression for surface excess, in which A is the surface area of the solution.

Since

$$n^{\alpha} \gg n^{\beta}$$

then

$$n^{\sigma} \approx n - n^{\alpha} = n - c_{\alpha}V_{\alpha} \tag{9.6}$$

thus

$$\Gamma \approx \frac{n - n^{\alpha}}{A} \tag{9.7}$$

The surface excess can be considered to be the difference of mole number between the interfacial phase of the unit surface area and the bulk solution of the same volume as the interfacial phase of the unit surface area.

It is noted that the location of s–s is very important, which influences the volume of the solution and therefore influences $c_{\alpha}V_{\alpha}$ and Γ. Now let us set the location of line s–s as follows.

Suppose the solution in the beaker has the surface of the unit area, then the height of the solution in the beaker will equal the volume of the solution.

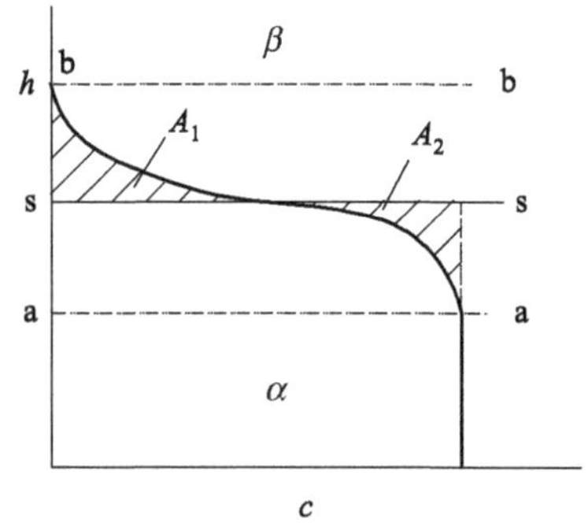

FIGURE 9.3 Location of line s–s.

Suppose the concentration of solvent is c, the height of the solution h, and the total mole number of the solvent n. Plotting the line of c versus h gives Fig. 9.3.

It can be seen that the concentration of solvent changes from the bottom of the beaker with the increase in the height of the solution. But it keeps constant below the line a–a, and starts to gradually decrease above line a–a, that is, in the interfacial region. When the height reaches line b–b, the concentration reaches zero. Therefore the total amount of solvent is given by

$$n = \int_0^{V_b} c\,dV = \int_0^{h_b} c\,dh = \text{the area below the curve}$$

Now let us place line s–s at a certain height so that the shaded area A_1 equals the shaded area A_2, thus the area below the curve will equal the area of the rectangle below line s–s.

Since

$$n^{\alpha} = c_{\alpha} V_{\alpha} = \text{the area of the rectangle}$$

then

$$n^{\sigma} = n - n^{\alpha} = 0$$

This leading to

$$\Gamma = 0$$

This implies that we set line s–s at the height that makes the surface excess of solvent equal to zero.

After setting line s–s, the surface excess of solute can be obtained by Eq. (9.5), in which n^{σ} = the area below the curve − the area of the rectangle, as show by the shaded area in Fig. 9.4.

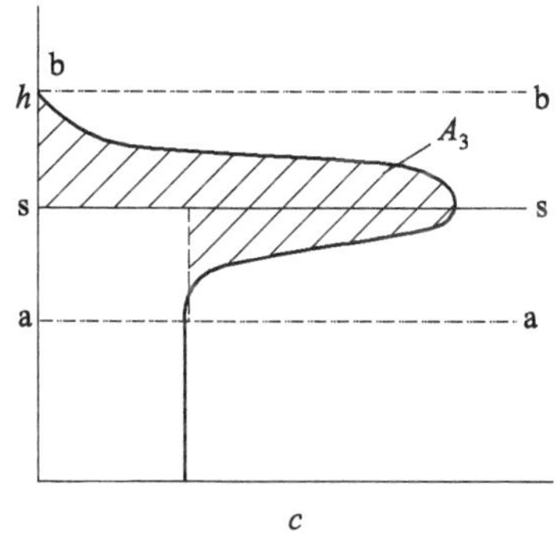

FIGURE 9.4 Surface excess of solute.

9.2.2 Gibbs Adsorption Equation

It is well known that the expression of the first law of thermodynamics is

$$dU = \delta Q - \delta W = TdS - \delta W \tag{9.8}$$

where U is the internal energy of the system, Q heat quantity, S entropy, and T thermodynamic temperature. For surface region, let us consider only the increase in surface area A and the change in components n_i, therefore, substitute surface work γdA for $-\delta W$, and add the changes of chemical potential caused by component changes, then the corresponding expression for surface region is given by

$$dU^\sigma = TdS^\sigma + \gamma dA + \sum_i \mu_i dn_i^\sigma \tag{9.9}$$

where i represents the component and μ the chemical potential. Integrating Eq. (9.9) at constant temperature and constant pressure from $A = 0$ to A gives

$$U^\sigma = TS^\sigma + \gamma A + \sum \mu_i n_i^\sigma \tag{9.10}$$

Then the differential form of Eq. (9.10) is given by

$$dU^\sigma = TdS^\sigma + S^\sigma dT + \gamma dA + Ad\gamma + \sum_i \mu_i dn_i^\sigma + \sum_i n_i^\sigma d\mu_i. \tag{9.11}$$

Comparing Eq. (9.11) with Eq. (9.9) gives

$$S^\sigma dT + Ad\gamma + \sum_i n_i^\sigma d\mu_i = 0 \tag{9.12}$$

At constant temperature

$$S^\sigma dT = 0$$

thus

$$-Ad\gamma = \sum_i n_i^\sigma d\mu_i \tag{9.13}$$

Dividing both sides of Eq. (9.13) by A gives

$$-d\gamma = \sum_i \Gamma_i d\mu_i \tag{9.14}$$

For the system of two components

$$-d\gamma = \Gamma_1 d\mu_1 + \Gamma_2 d\mu_2 \tag{9.15}$$

where subscript 1 represents solvent and subscript 2 solute. As described earlier, the surface excess of solvent equals zero; therefore, Eq. (9.15) becomes

$$-d\gamma = \Gamma_2 d\mu_2 \tag{9.16}$$

The chemical potential is related to concentration by

$$\mu = \mu^\ominus + RT \cdot \ln c$$

thus

$$d\mu_2 = RT \cdot d\ln c_2 \tag{9.17}$$

Substituting Eq. (9.17) for $d\mu_2$ in Eq. (9.16) gives

$$-d\gamma = \Gamma_2 RT \cdot d\ln c_2$$

therefore

$$\Gamma_2 = -\frac{d\gamma}{RT \cdot d\ln c_2} = -\frac{d\gamma}{RT(dc_2/c_2)} = -\frac{c_2}{RT}\frac{d\gamma}{dc_2} = -\frac{c_2}{RT}\left(\frac{\partial\gamma}{\partial c_2}\right)_T \tag{9.18}$$

This is the form in which the Gibbs equation is usually quoted. It can be seen that if the surface tension decreases with the increase in the concentration of solute, that is, $\partial\gamma/\partial c < 0$, the surface excess will be positive, that is, the positive adsorption. If the surface tension increases with the increase in the concentration of solute, that is, $\partial\gamma/\partial c > 0$, the surface excess will be negative, that is, the negative adsorption.

Example 9.1 Surface Excess of Solution

By using the device designed by McBain and his students, researchers scraped off a thin surface layer of a nonionic surfactant solution at 25°C. The area and the volume of this surface layer are 300 cm^2 and 2 cm^3, respectively. It contains 4.013×10^{-5} mol surfactant, whereas the bulk solution with the same volume as the surface layer contains 4.000×10^{-5} mol surfactant. Calculate the surface tension of this solution.

Solution

Suppose the surface tension of the solution linearly decreases with the increase in the concentration of surfactant. The Gibbs equation can be expressed as

$$\Gamma_2 = -\frac{c_2}{RT}\left(\frac{\partial \gamma}{\partial c_2}\right)_T = -\frac{c_2}{RT}\cdot\frac{\gamma-\gamma_0}{c_2-0} = -\frac{\gamma-71.97\times10^{-3}}{RT}\ (\text{mol m}^{-2})$$

or

$$\Gamma_2 = \frac{4.013\times10^{-5}-4.000\times10^{-5}}{300\times10^{-4}}\ (\text{mol m}^{-2})$$

thus

$$\frac{4.013\times10^{-5}-4.000\times10^{-5}}{300\times10^{-4}} = -\frac{\gamma-71.97\times10^{-3}}{8.314\times298}$$
$$\gamma = 61.27\ (\text{mN m}^{-1})$$

Example 9.2 Surface Excess of Solution

The surface tension of the aqueous solution of ethanol obeys expression $\gamma = 72 - 0.5c + 0.2c^2$ at 25°C, where c is the concentration of ethanol. Calculate the surface excess of ethanol at this temperature.

Solution

Since the surface tension obeys

$$\gamma = 72 - 0.5c + 0.2c^2$$

then

$$\frac{\partial \gamma}{\partial c} = -0.5 + 0.4\times0.5 = -0.3$$
$$\Gamma_2 = -\frac{c_2}{RT}\left(\frac{\partial \gamma}{\partial c_2}\right)_T = -\frac{0.50}{8.314\times298}\times(-0.3) = 6.05\times10^{-5}\ (\text{mol m}^{-2})$$
$$= 6.05\times10^{-9}\ (\text{mol cm}^{-2})$$

9.2.3 Adsorption at Surface of Solution

Line C in Fig. 9.1 shows that when the concentration of surfactant is low, a linear relation between γ and c can be obtained

$$\gamma = \gamma_0 - Bc \tag{9.19}$$

where B is the slope. Then we have

$$\gamma_0 - \gamma = Bc \tag{9.20}$$

Combining Eq. (8.41) with Eq. (9.20) gives

$$\pi = Bc$$

thus

$$\Gamma_2 = -\frac{c_2}{RT}\left(\frac{\partial \gamma}{\partial c_2}\right)_T = -\frac{c_2}{RT}\cdot\frac{\gamma-\gamma_0}{c_2-0} = \frac{Bc}{RT} = \frac{\pi}{RT} \tag{9.21}$$

Γ_2 is defined as the difference of mole numbers between the interfacial phase of the unit surface area and the same volume of bulk solution. If Γ_2 is considered as approximately the total mole number of solute in the surface layer of the unit surface area, then the surface area occupied by one mole of solute is given by

$$A = \frac{1}{\Gamma_2} \tag{9.22}$$

Combining Eq. (9.22) with Eq. (9.21) gives

$$\frac{1}{A} = \frac{\pi}{RT}$$

that is,

$$\pi A = RT \tag{9.23}$$

Dividing both sides of Eq. (9.23) by the Avogadro constant gives

$$\pi a = K_B T \tag{9.24}$$

where a is the area occupied by one molecule and K_B the Boltzmann constant. Eq. (9.23) and Eq. (9.24) are the expressions of two-dimensional state equations of ideal gas.

If the concentration of surfactant is continually increased to the nonlinear region of line C in Fig. 9.1, substituting the differential Szyszkowski empirical equation [Eq. (9.2)] into the Gibbs equation will give

$$\Gamma_2 = \frac{\Gamma_m c}{a+c} \tag{9.25}$$

where

$$\Gamma_m = \frac{b\gamma_0}{RT}$$

Eq. (9.25) is the expression for adsorption isotherm of solution. If the concentration of surfactant is increased to very high values, Eq. (9.25) will become

$$\Gamma = \Gamma_m \tag{9.26}$$

If the concentration of surfactant is very small, Eq. (9.25) will become the linear relation as shown in Eq. (9.21).

9.3 SURFACTANT

9.3.1 Characteristics, Chemical Structure, and Classification of Surfactant

The substance which is able to lower the surface tension of water remarkably is termed a surfactant, as shown by curve C in Fig. 9.1. Since $\partial\gamma/\partial c_2 < 0$ for surfactants, their surface excess at the surface of water is positive according to the Gibbs equation, which means positive adsorption. Surfactants are widely used in practice, for example, emulsification, breaking of emulsion, foaming, defoaming, dispersion, flocculation, increasing of solubility, wetting, scale inhibition, and corrosion inhibition. As far as the chemical structure is concerned, a surfactant molecule has both a hydrophilic polar group and a hydrophobic nonpolar hydrocarbon chain on one molecule, as shown in Fig. 9.5, for example, CH_3—$(CH_2)_n$—OH (alcohol), CH_3—$(CH_2)_n$—COOH (carboxylic acid), and CH_3—$(CH_2)_n$—NH_2 (amine). Therefore, the molecule of the surfactant is always adsorbed at the surface of water, with the hydrophilic groups oriented toward the water phase and the hydrophobic hydrocarbon chains oriented away from the water phase. Due to the hydrophobic effect of the hydrocarbon chain, the oriented molecule of the surfactant can reduce the vertically inward composite force, thus decreasing the surface tension. The more molecules of the surfactant in the surface layer, the greater the decrease in surface tension. In the beginning, the surfactant added to water is less, thus the concentration is lower; the molecules of the surfactant may roughly lie flat on the surface of water but still have the orientation to some extent due to the repulsion between the hydrophobic chains and water as well as the attraction between polar groups and water. The situation is illustrated in Fig. 9.6a. With the increase in concentration, the hydrocarbon chains become more and more oriented toward the air, as shown in Fig. 9.6b. When the concentration reaches very high, the molecules

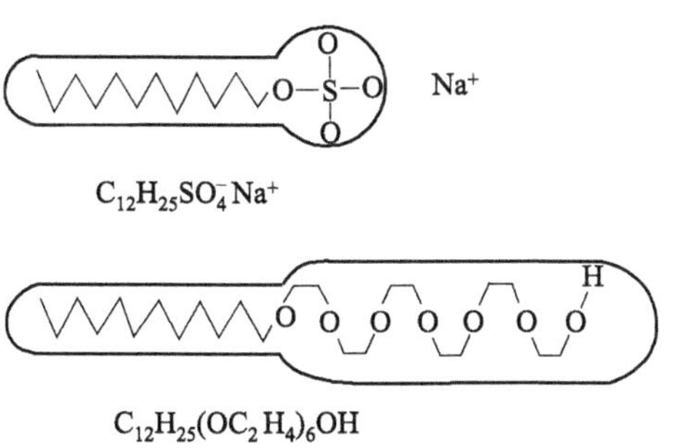

$C_{12}H_{25}SO_4^-Na^+$

$C_{12}H_{25}(OC_2H_4)_6OH$

FIGURE 9.5 Molecular structure of surfactant.

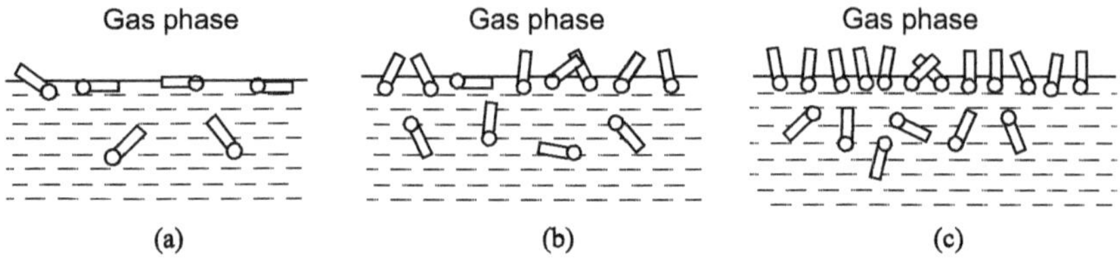

FIGURE 9.6 Orientation of surfactant molecules at interface. (a) the concentration of surfactant is low; (b) the concentration of surfactant is increased; (c) the concentration of surfactant reaches very high.

arrange "side by side" at the surface, reaching the adsorption saturation, as shown in Fig. 9.6c.

Surfactants are classified as anionic, cationic, and nonionic according to the charge carried by the surface-active part of the molecule.

Example 9.3 Surfactant

Organic corrosion inhibitors are widely used in industrial circulating cooling water, for example, organic amine and sodium gluconate. Give the acting mechanism of the organic corrosion inhibitors.

Solution

Commonly, the organic corrosion inhibitors used in industrial circulating cooling water are surfactants. They adsorb on the surface of metal with their polar groups, thus changing the structure of the electric double layer and increasing the activation energy for metal ionization. In addition, their nonpolar groups orient toward the water phase, forming a layer of hydrophobic film, thus preventing corrosion reactants and corrosion products from diffusion; as a result, the corrosion is inhibited.

9.3.2 Formation of Micelles

Fig. 9.7 is presented as a typical illustration of the way surface tension γ varies with the concentration of surfactant solute in aqueous solution.

In some instances, it is noted that there is a minimum at the curve, as shown by the dotted line. Two questions were proposed for this kind of curve in the past. One is that when the concentration of surfactant increases to a certain value, the surface tension will not change anymore, that is, $\partial\gamma/\partial c_2 \approx 0$. Does this imply the positive adsorption becomes zero adsorption? The other one is that after reaching the minimum value the surface tension rises with the concentration of surfactant, for example, sodium lauryl sulfate. Does this imply that the positive adsorption abruptly becomes a negative adsorption? The answer is no, both of them are not possible in practice.

McBain pointed out that this seemingly anomalous behavior could be explained in terms of organized aggregates, or micelles, of the surfactant ions in

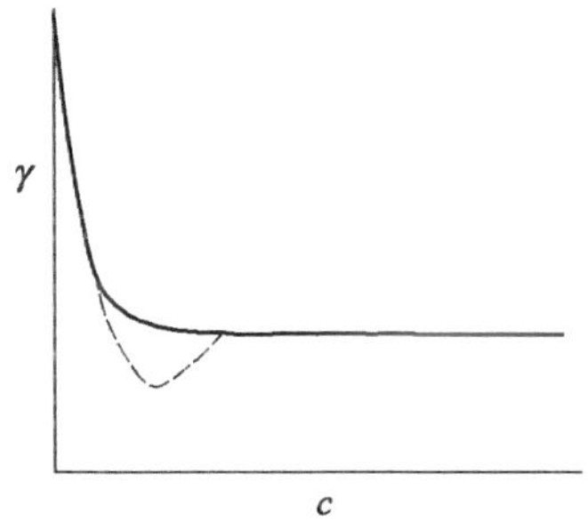

FIGURE 9.7 **Variation of γ with c in aqueous solution.**

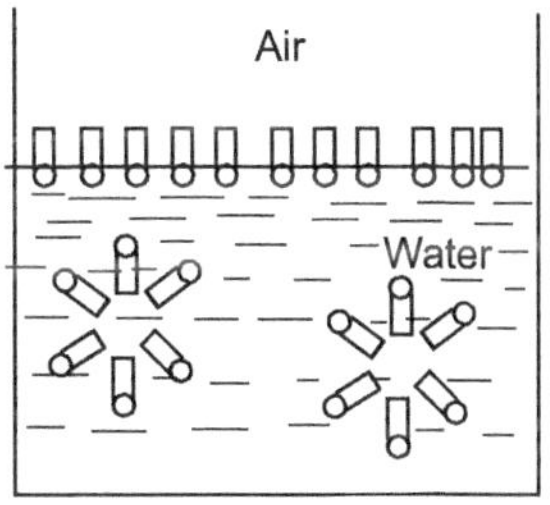

FIGURE 9.8 **Formation of micelle.**

which the lipophilic hydrocarbon chains are oriented toward the interior of the micelle, leaving the hydrophilic groups in contact with the aqueous medium, thus forming the thermodynamically stable system. Fig. 9.8 schematically shows the formation of the micelle. When added to water, at first the surfactant molecules enter the surface layer preferentially to reduce the surface tension as much as possible. When the concentration reaches a certain value and the surface layer becomes saturated, they will enter the interior of the solution, and the micelles begin to form because of the lipophilic property of the hydrocarbon chains. The previously mentioned concentration in which micelle formation becomes appreciable is termed the critical micelle concentration (c.m.c.). Commonly they are in the range of 0.001~0.02 mol dm^{-1}. Many nonionic surfactants also form micelles, often at very small concentrations (c. 10^{-4} mol dm^{-1}).

The aggregate number of micelle is commonly in the range of 100 ± 50. Thermal agitation and electrostatic repulsion between the charged head groups on the surface of the micelle oppose this aggregating tendency. Consequently, a low c.m.c. would be expected to be favored by

1. Increasing the hydrophobic part of the surfactant molecules
2. Lowering the temperature
3. The addition of simple salt which reduces the previously mentioned repulsion force by their screening action

Now we can explain the questions proposed earlier. When the concentration of surfactant reaches c.m.c., the micelles will begin to form; therefore, the number of single molecules in the solution will not continually increase, thus the surface tension will not decrease significantly and the adsorption will keep constant. In addition, sodium lauryl sulfate commonly contains some dodecanol which has very strong surface activity; therefore, dodecanol is adsorbed on the surface of the solution strongly, decreasing the surface tension significantly, but once the micelles form, they will be desorbed and enter the micelles (solubilization), thus resulting in the rise up of surface tension to some extent.

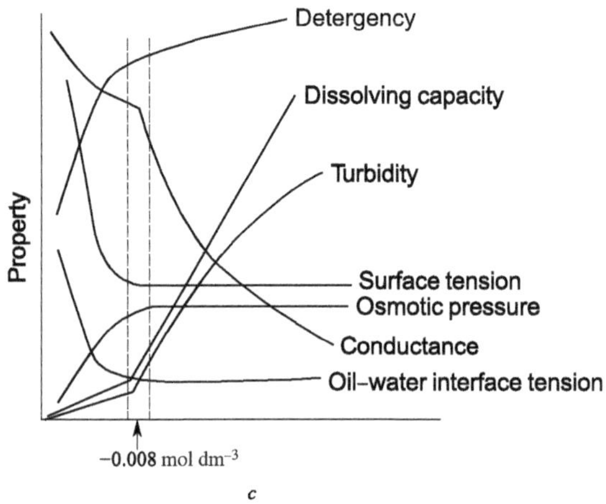

FIGURE 9.9 Physical properties of sodium dodecyl sulfate solution at 25°C.

9.3.3 Effect of Micelles on the Property of Solution

Solutions of highly surface-active materials exhibit unusual physical properties. In dilute solutions they act as normal electrolytes, but at the concentration above c.m.c. the abrupt changes in several physical properties, mainly in colligative properties, such as osmotic pressure, conductance, turbidity, dissolving capacity, detergency, and surface tension, take place as shown in Fig. 9.9. The reason is that the surfactant exists mainly in the form of micelles above c.m.c. The micelle is much bigger than the single molecule; therefore, the increase in particle number with the increase in concentration becomes slower.

It can be seen that solubilization takes place near c.m.c. For example, the solubility of benzene in water is only 0.07 g per 100 g water, but it reaches 7 g per 100 g water in 10% sodium oleate aqueous solution. The reason is that the lipophilic hydrocarbon chains are oriented toward the interior of the micelle, thus forming a small "organic phase" in the center of the micelle, which can dissolve some insoluble organics. As a result, the solubility of the organics in water is significantly increased. In addition, it is clear that by using the solubilization we can enhance the detergency greatly.

9.3.4 Dependence of Surfactant Solubility on Temperature

1. Krafft temperature of ionic surfactant
   Ionic surfactants exhibit another unusual phenomenon in that their solubilities show a rapid increase above a certain temperature, known as the Krafft point, as shown in Fig. 9.10. The explanation of this behavior arises from the fact that the unassociated surfactant has a limited solubility, whereas the micelles are highly soluble. Below the Krafft temperature the solubility of

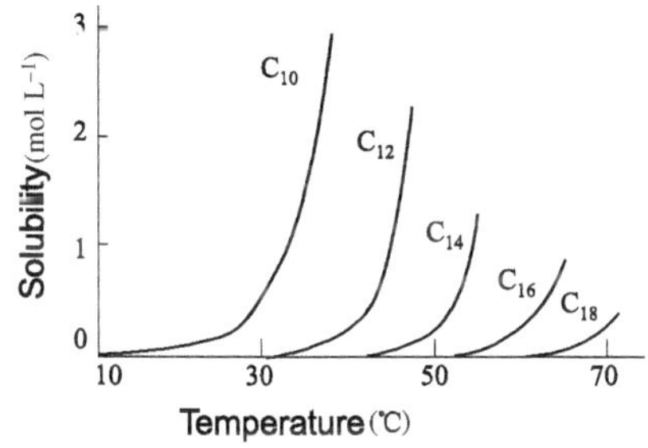

FIGURE 9.10 **Relation between solubility and temperature for alkylbenzene sulfonate.**

the surfactant is insufficient for micellization. As the temperature is raised the solubility slowly increases until, at the Krafft temperature, the c.m.c. is reached. A relatively large amount of surfactant can now be dispersed in the form of micelles, so that a large increase in solubility is observed.

2. Cloud point of nonionic surfactant
As the temperature is raised to a certain value, some surfactant solution will become turbid, that is, the solubility of the surfactant will become smaller. The temperature at which the solution becomes turbid is termed "cloud point." For example, the molecule of polyoxyethylene surfactant dissolved in water reacts with water molecules, forming a hydrogen bond, but when the temperature is raised to the "cloud point," the hydrogen bond will be broken, thus the surfactant separates out. As a result, the solution becomes turbid. Generally speaking, the more hydrophilic the nonionic surfactant, the higher the "cloud point." Therefore, the hydrophilic or hydrophobic properties of a surfactant can be measured in terms of the value of its "cloud point."

9.3.5 Dependence of Surfactant Properties on Its Structure

The more hydrophilic the polar group of the surfactant molecule, the more hydrophilic the surfactant. The hydrophilic properties of the common polar groups decrease in the following order:

$$-SO_4^- > -COO^- > -SO_3^- > NR_4^+ > -COO- > -COOH > OH > -O-$$
$$> -CH_2-CH_2O-$$

The strong-hydrophilic surfactants are applicable for washing and often used as detergents, such as sodium dodecylbenzenesulfonate. The weak-hydrophilic surfactants are applicable for defoaming and often used as antifoaming agents, such as fatty alcohol and ether. The nonionic surfactant synthesized by epoxyethane and propylene epoxide is a very efficient antifoaming agent, which can make the liquid films between bubbles become thinner, thus causing the rupture of liquid film, and then the destruction of foam.

The more hydrophobic the nonpolar group of surfactant molecule, the more hydrophobic the surfactant. The hydrophobic properties of the common nonpolar groups decrease in following order:

Alkyl > alkylene > aryl radical with alkyl chain > aryl radical
$> -CH_2-CH_2-CH_2O-$

The surfactant with a nonpolar group similar to oil in structure is applicable to emulsify oil; for example, the aromatic dye can be emulsified by using the surfactant with aryl radical, whereas mineral oil can be emulsified by using the surfactant with fatty chains.

The geometric position of polar group or nonpolar group on the surfactant molecule also has an effect on its properties. If the hydrophilic group is on the end of the molecule, the surfactant will be a strong detergent; if the hydrophilic group is on the middle of the molecule, the surfactant will be a strong wetting agent and a good penetrating agent. In addition, if the molecule has a branch chain, the surfactant will also have strong wetting effect and good permeability.

Anionic surfactants produce lime soap in hard water and may separate out or decompose in acid solution. Cationic surfactants are not influenced by hardness of water and acid medium. The fourth amine salt can be used for disinfection. Nonionic surfactants are not influenced by hardness and pH value. Amphoteric surfactants are anionic in alkaline solution and are cationic in acid solution.

Chapter 10

Surface of Solids

Chapter Outline

Colloid and Interface Chemistry for Water Quality Control. http://dx.doi.org/10.1016/B978-0-12-809315-3.00010-4

10.1 BASIC PRINCIPLES

The difference between a solid surface and a liquid surface is that the molecules on the liquid surface are movable; therefore, the liquid surface is smooth, uniform, and contracts due to the effect of surface tension, whereas the molecules on the solid surface are not movable; therefore, the solid surface is uneven and cannot contract. The similarity between a solid surface and a liquid surface is that there are imbalances of attractive forces at the surfaces of both of them, which cause surface free energy. The surface free energy of a solid surface results in the following phenomena.

1. Adsorption
   A solid surface lowers its surface free energy by adsorption rather than decreasing the surface area like liquid surface. The reason is that the molecule adsorbed at the surface can offset the vertically inward composite force to some extent. As a result, when a gas or vapor or solute is brought into contact with a clean solid surface some of it will become attached to the surface in the form of an adsorbed layer. Here the solid is generally referred to as the adsorbent, and the substances adsorbed at the surface as the adsorbate.
2. Enhanced vapor pressure of small particles
   Applying Kelvin equation [Eq. (8.18)] to small particles gives

$$RT \cdot \ln\frac{P_r}{p_0} = \frac{2\gamma M}{r\rho} = \frac{2\gamma V}{r} \tag{10.1}$$

   where P_r is the vapor pressure of small particles at equilibrium; P_0 the vapor pressure of solid of flat surface at equilibrium. It shows that the smaller the particles, the higher the vapor pressure; therefore, if a solid material is divided into fine particles, its melting point will be lowered.
3. Enhanced solubility of small particles
   Again applying Kelvin equation [Eq. (8.18)] to small particles gives

$$RT \cdot \ln\frac{x_r}{x_0} = \frac{2\gamma M}{r\rho} = \frac{2\gamma V}{r} \tag{10.2}$$

   where x_r is the solubility of small particles at equilibrium; x_0 the solubility of solid of flat surface at equilibrium. It shows that the smaller the particle, the higher their solubility. Therefore, it will be difficult for an insoluble material to precipitate from solution if there are no larger particles as the crystal nucleus in the solution, and it will be easy to form a saturated solution.

10.1.1 Physical Adsorption and Chemical Adsorption

Adsorptions of gas on solid surfaces are classified as physical adsorption and chemical adsorption. The forces involved in the adsorption of gas and vapor by

TABLE 10.1 Differences Between Physical Adsorption and Chemical Adsorption

Property	Physical adsorption	Chemical adsorption
Force of adsorption	Van der Waals force	Chemical bond
Heat of adsorption	Similar to liquefaction	Similar to chemical reaction
Selectivity	No	Yes
Reversibility	Reversible	Irreversible
Adsorption rate	Quick, no activation energy	Slow, requires activation energy
Adsorption layer	Monolayer or multilayer	Monolayer
Pressure for monolayer	$P/P_0 \approx 0.1$	$P/P_0 \ll 0.1$
Temperature	Near or below boiling point	Much higher than boiling point

solids may be van der Waals forces, similar to the force involved in liquefaction, or strong forces, such as those which are operative in the formation of chemical bonds. The former are responsible for physical adsorption and the latter for chemical adsorption. The difference between physical adsorption and chemical adsorption are listed in Table 10.1. Although there are many differences between physical adsorption and chemical adsorption, it is more obvious to distinguish them by reversibility and the heat of adsorption.

10.1.2 Thermodynamics of Adsorption

1. Thermodynamics factors influencing adsorption
   Thermodynamics of adsorption is the study on adsorption equilibrium. The factors which influence adsorption equilibrium are temperature, solubility of adsorbate, polarity, pore size of adsorbent, and more.
   a. Temperature: Generally speaking, adsorption is an exothermic process. The extent of adsorption (at equilibrium condition), therefore, increases with decreasing temperature. For some substances with very limited solubility, increase in temperature will lead to the increase in solubility, thus leading to the increase in the extent of adsorption.
   b. Solubility: The smaller the solubility of adsorbate, the more easily it is adsorbed, thus, the higher the extent of adsorption.
   c. Polarity: Polar adsorbent adsorbs polar solute more easily, and nonpolar adsorbent adsorbs nonpolar solute more easily. Polar solute is adsorbed from nonpolar solvent easily, and nonpolar solute is adsorbed from polar solvent easily. Figs. 10.1–10.3 illustrate the effect of polarity on adsorption.

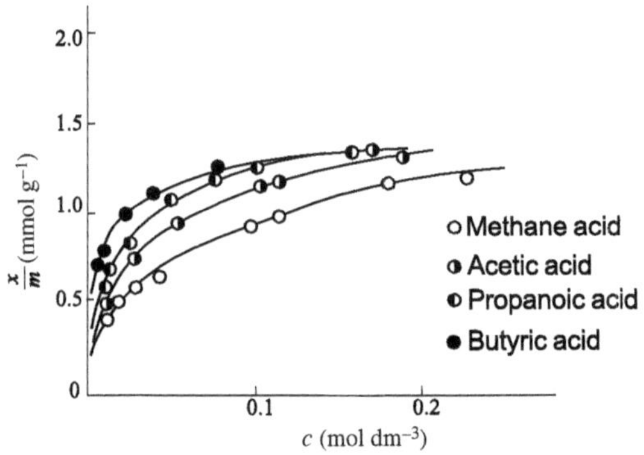

FIGURE 10.1 Fatty acids are adsorbed by activated sugar charcoal from aqueous solution (400°C).

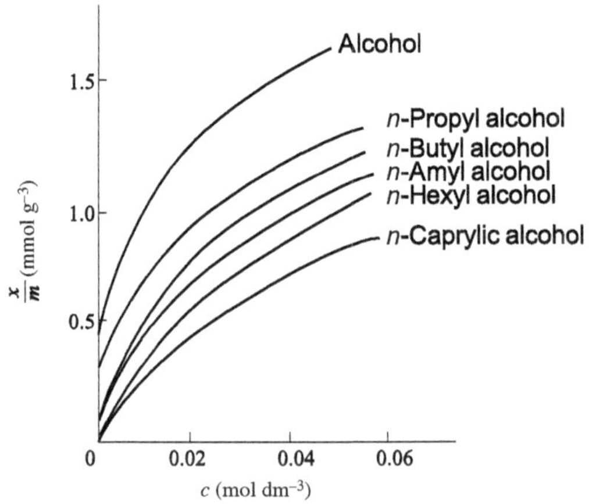

FIGURE 10.2 Fatty alcohols are adsorbed by silica gel from carbon tetrachloride solution (400°C).

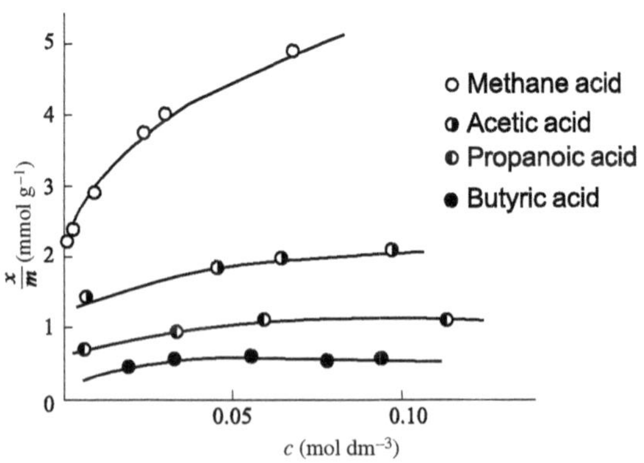

FIGURE 10.3 Fatty acids are adsorbed by activated sugar charcoal from carbon tetrachloride solution (400°C).

Fig. 10.1 shows fatty acids are adsorbed by activated sugar charcoal. The activated carbon is known as a nonpolar adsorbent; on the other hand, the longer the carbon chain of fatty acids, the stronger its nonpolar property. Therefore, the fatty acids with longer carbon chains are easily adsorbed to the activated sugar charcoal from water which is a polar solvent. Fig. 10.2 shows fatty alcohols are adsorbed by silica gel. Silica gel is known as a polar adsorbent; on the other hand, the shorter the carbon chain of fatty alcohol, the stronger its polar property. Therefore, the fatty alcohols with shorter carbon chains are easily adsorbed to silica gel from carbon tetrachloride which is a nonpolar solvent. Fig. 10.3 shows fatty acids are adsorbed by activated sugar charcoal from carbon tetrachloride solution. Although activated carbon is a nonpolar adsorbent relative to water, it is also a polar adsorbent relative to carbon tetrachloride; therefore, the adsorption extent of fatty acids with a shorter carbon chain, that is, the stronger polarity, are greater.

d. Pore size: Pore size influences not only the rate of adsorption but also the equilibrium of adsorption. If the pore size is relatively small, only the solutes with shorter hydrocarbon chains can enter the pores easily, leading to the greater extent of adsorption, whereas the solutes with longer hydrocarbon chains cannot enter the pores easily, leading to the smaller extent of adsorption.

2. Heat of adsorption

According to the thermodynamic principle, at constant temperature

$$\Delta G = \Delta H - T\Delta S$$

where G is free energy, H enthalpy, and S entropy. With a spontaneous process, the adsorption results in a decrease in free energy

$$\Delta G < 0$$

When adsorption takes place, the gas molecules are restricted to two-dimensional motion. Gas adsorption processes are, therefore, accompanied by a decrease in entropy, that is,

$$\Delta S < 0$$

thus

$$-T\Delta S > 0$$

It is evident that ΔH must be negative, that is, the adsorption of gases and vapors on solids is always an exothermic process. However, this conclusion is applicable only for inert adsorbents. For chemical adsorption, since the surface atoms may be chemically changed, the ΔS may be positive; for the adsorption from solution, since the adsorption is accompanied by desolvation, the ΔS is frequently positive.

Heats of adsorption can be measured by direct calorimetric method. Isosteric heats of adsorption can be calculated from the Clausius–Clapeyron equation, which is the expression for the gas–liquid equilibrium

$$\left(\frac{\partial \ln P}{\partial T}\right)_V = \frac{-\Delta H}{RT^2} = \frac{Q}{RT^2} \tag{10.3}$$

where ΔH is the heat effect of adsorption, and $Q = -\Delta H$, which is isosteric heat. Integrating Eq. (10.3) gives

$$\Delta H = \frac{RT_1T_2}{T_2 - T_1} \cdot \ln \frac{p_1}{p_2} \tag{10.4}$$

It can be seen that if we substitute two different temperatures and the two corresponding pressures at equilibrium into Eq. (10.4), we will obtain the heat of adsorption.

10.2 ADSORPTION OF GAS AT A SOLID SURFACE

10.2.1 Adsorption Isotherm

Amount of adsorption is defined as the amount of solute adsorbed by unit mass of adsorbent. Generally speaking, the amount of adsorption is the function of the temperature and the gas pressure at equilibrium

$$V = f(P,T) \tag{10.5}$$

where V is the amount of adsorption which can be measured in terms of the volume of the gas adsorbed (transformed as the value on standard condition), P the gas pressure, and T the thermodynamic temperature. If the temperature remains unchanged during the adsorption process, the amount of adsorption is only the function of equilibrium gas pressure, that is,

$$V = f(P) \tag{10.6}$$

The empirical or theoretical representation for the amount of adsorption as a function of the equilibrium gas pressure is called the adsorption isotherm. Brunauer has classified adsorption isotherms into five characteristic types, which are shown in Fig. 10.4.

Type I isotherms show a fairly rapid rise in the amount of adsorption, with pressure up to a limiting value. They are referred to as Langmuir-type isotherms and obtained when adsorption is restricted to a monolayer. Chemical adsorption isotherms, therefore, approximate to this shape. Type I isotherms have also been found for physical adsorption on solids containing a very fine pore structure. Type II isotherms are frequently encountered and represent multilayer physical adsorption on nonporous solids. Physical adsorption on

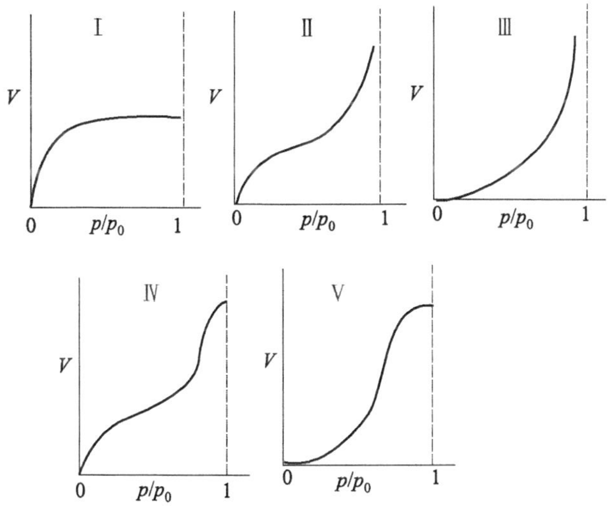

FIGURE 10.4 **Basic types of adsorption isotherms.**

microporous solids can also result in Type II isotherms. Type IV isotherms level off near the saturation vapor pressure and are considered to reflect capillary condensation in porous solids, the effective pore diameters generally being between about 2 and 20 nm. The upper limit of adsorption is mainly governed by the total pore volume. Type III and V show no rapid initial uptake of gas and occur when the forces of adsorption in the first monolayer are relatively small. These isotherms are rare.

10.2.2 Langmuir Adsorption Isotherm Equation

The first quantitative theory of the adsorption of gases was given in 1916 by Irving Langmuir as follows. Suppose μ is the number of gas molecules which strike a solid surface of 1 $cm^2\ s^{-1}$, and among them $a\mu$ gas molecules are condensed (a is the ratio coefficient); again v is the number of molecules which evaporate from the solid surface of 1 $cm^2\ s^{-1}$. Let the concentration of the molecules adsorbed at the surface be Γ (cm^{-2}). Then the adsorption rate is given by

$$\frac{d\Gamma}{dt} = a\mu - v \tag{10.7}$$

When the adsorption reaches equilibrium

$$\frac{d\Gamma}{dt} = 0$$

thus

$$a\mu = \nu$$

If the released heat is q when one molecule is condensed, then the molecules with the energy above q will be able to leave the surface. Their fraction in the total molecules adsorbed at the surface of 1 cm^2 can be calculated from the Boltzmann distribution, thus we have

$$\nu = k_0 e^{-q/K_B T} \tag{10.8}$$

where k_0 is the ratio coefficient. According to kinetic molecular theory

$$\mu = \frac{P}{(2\pi m K_B T)^{1/2}} \tag{10.9}$$

where P is the gas pressure and m the molecular mass.

Let us make the following assumptions: (1) there is no interaction between molecules adsorbed at the surface; (2) the surface is uniform everywhere; and (3) only monomolecular layer adsorption takes place. Based on these assumptions the Langmuir adsorption isotherm equation is deduced as follows.

If θ is the fraction of the surface area covered by adsorbed molecules at any time, the rate of evaporation of molecules from the surface will be proportional to θ and given by

$$\nu = \nu_1 \theta \tag{10.10}$$

where ν_1 is the rate of evaporation of the molecules from the surface when $\theta = 1$. The rate of condensation of molecules on the surface is proportional to the fraction of surface that is bare, $1 - \theta$, and to the rate at which molecules strike the surface at a given temperature. The rate of condensation is, therefore, set equal to

$$a\mu = a_0 (1-\theta)\mu \tag{10.11}$$

where a_0 is the ratio coefficient when $\theta = 0$. At equilibrium, the rate of condensation equals the rate of evaporation

$$a_0 (1-\theta)\mu = \nu_1 \theta \tag{10.12}$$

thus

$$\theta = \frac{\frac{a_0}{\nu_1}\mu}{1+\frac{a_0}{\nu_1}\mu} \tag{10.13}$$

Substituting Eq. (10.8) and Eq. (10.9) into Eq. (10.13) gives

$$\theta = \frac{\dfrac{a_0}{k_0 e^{-q/K_BT}(2\pi m K_B T)^{1/2}} \cdot P}{1 + \dfrac{a_0}{k_0 e^{-q/kT}(2\pi m K_B T)^{1/2}} \cdot P} \tag{10.14}$$

Suppose

$$b = \frac{a_0}{k_0 e^{-q/K_BT}(2\pi m K_B T)^{1/2}} \tag{10.15}$$

Eq. (10.14) will become

$$\theta = \frac{bP}{1+bP} \tag{10.16}$$

Eq. (10.16) is referred to as the Langmuir adsorption equation, where b is called the adsorption coefficient. It can be seen from Eq. (10.15) that the more the released heat, the greater the value of b. Therefore the value of b reflects the adsorption strength. The greater the value of b, the stronger the adsorption.

Let the amount of adsorption be V and the amount of adsorption at monolayer coverage (or the adsorption capacity) be V_m, then θ can be expressed in terms of their ratio

$$\frac{V}{V_m} = \frac{bP}{1+bP} \tag{10.17}$$

or

$$V = \frac{V_m bP}{1+bP} \tag{10.18}$$

which is the commonly used form of the Langmuir equation. If b or P is very small, Eq. (10.18) will be simplified as

$$V = V_m bP \tag{10.19}$$

If b or P is great, Eq. (10.18) will be transformed as

$$V = V_m \tag{10.20}$$

The linear equation of Langmuir adsorption isotherm can be obtained by appropriate transformation of Eq. (10.18) as follows

$$\frac{P}{V} = \frac{1}{V_m b} + \frac{P}{V_m} \tag{10.21}$$

Based on Eq. (10.21) and experiment data, plotting the straight line of $\frac{P}{V}$ versus P gives

$$\text{Slope} = \frac{1}{V_m} \tag{10.22}$$

$$\text{Intercept} = \frac{1}{V_m b} \tag{10.23}$$

Combining Eqs. (10.22) and (10.23) gives constant b and constant V_m. Alternatively we can make a linear fitting by Excel to obtain the slope, the intercept, as well as the multiple correlation coefficient R^2, which can be used to determine the degree of fitting, thus judge if the data obtained in the experiment conforms to the Langmuir equation.

In some cases, the data obtained in the experiment do not conform very well to the Langmuir equation. The reason is that some assumptions which are not in accord with practice were made in the derivation of the equation. Actually, there are interactions between the adjacent molecules adsorbed at the surface. With adsorption proceeding, the concentration of molecules of the adsorbate at the surface increases, leading to the increase in the interaction between molecules adsorbed, thus decreasing the tendency of adsorption and increasing the tendency of desorption. In addition, the surface is not uniform everywhere. At the initial stage of adsorption, the degree of surface coverage is lower and the molecules are adsorbed preferentially on the site of higher activity; therefore, adsorption takes place easily, but desorption does not take place easily. With the adsorption proceeding, the coverage gradually increases and the bare sites of higher activity become less and less; therefore, the adsorption becomes more and more difficult and the desorption becomes easier and easier. In summary, the linear expression of Eqs. (10.10) and (10.11) proposed in derivation due to the basic assumptions are not very exact, thus leading to the difference between the result predicted by the Langmuir equation and the practical experimental result.

To solve the problem caused by the inhomogeneity of practical surface, we can divide the surface into many types (represented by the different values of i) according to the inhomogeneity, then the corresponding amount of adsorption in saturation and adsorption coefficient for each type is $V_{m,i}$ and b_i, respectively. Langmuir gave the expression as follows.

$$V = \sum_i \frac{V_{m,i} b_i P}{1 + b_i P} \tag{10.24}$$

The sum of all $V_{m,i}$ should be equal to V. Thus the difference between the result predicted by the Langmuir equation and the practical result can be reduced to some extent by this method.

When the different components of a gas mixture compete for the adsorption sites on a solid surface the Langmuir equation takes the general form

$$\frac{V_i}{V_{m,i}} = \frac{b_i P_i}{1+\sum b_i p_i} \quad (10.25)$$

Example 10.1 Langmuir Adsorption Isotherm Equation

The volume of nitrogen gas adsorbed per gram of activated carbon (transformed as the volume at standard state) at 0°C varies with the pressure as shown in the following table.

P (Pa)	57.2	161	523	1728	3053	4527	7484	10310
V (mL g^{-1})	0.111	0.298	0.987	3.043	5.082	7.047	10.31	13.05

Give the expression of the Langmuir adsorption isotherm equation according to the experiment data.

Solution

Use the linear expression of the Langmuir adsorption isotherm equation

$$\frac{P}{V} = \frac{1}{V_m b} + \frac{P}{V_m}$$

and transform the data into the following form.

P (Pa)	57.2	161	523	1728	3053	4527	7484	10310
P/V (Pa g mL^{-1})	515	540	530	568	601	642	726	790

Making a linear regression by using Excel gives the straight line of $\frac{P}{V}$ versus P as follows

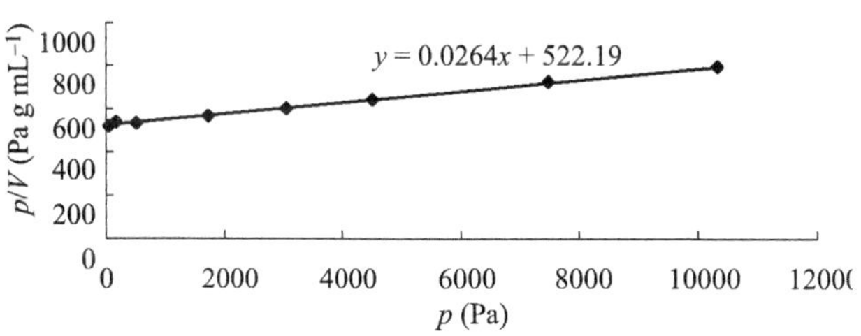

From this straight line we obtain

$$\text{Intercept} = 522 \text{ Pa g cm}^{-3} = 1/(V_m b)$$

$$\text{Slope} = 0.0264 \text{ g cm}^{-3} = 1/V_m$$

Therefore

$$V_m = 37.8\,\text{cm}^3\,\text{g}^{-1},\ b = 5.07\times10^{-5}\ \text{Pa}$$

Then the expression of the Langmuir adsorption isotherm equation is given by

$$V = \frac{bV_mP}{1+bP} = \frac{1.91\times10^{-3}P}{1+5.07\times10^{-5}P}$$

Example 10.2 Langmuir Adsorption Isotherm Equation of Gas Mixture

A gas mixture consists of component A and component B. Derive the expression of the Langmuir adsorption isotherm equation for this gas mixture.

Solution

According to Eq. (10.12)

$$\text{Adsorption rate of A is } r_a = k_aP_A(1-\theta_A-\theta_B)$$

$$\text{Desorption rate of A is } r_d = k_d\theta_A$$

where k_a and k_d are adsorption rate constant and desorption rate constant, respectively. At equilibrium $r_a = r_d$, that is,

$$k_d\theta_A = k_aP_A(1-\theta_A-\theta_B)$$

Divide both sides of this equation by k_d, and let $b_A = \frac{k_a}{k_d}$, then we get

$$\frac{\theta_A}{1-\theta_A-\theta_B} = b_AP_A \quad \text{(i)}$$

Similarly we get

$$\frac{\theta_B}{1-\theta_A-\theta_B} = b_BP_B \quad \text{(ii)}$$

Combining Eqs. (i) and (ii) gives

$$\theta_A = \frac{b_AP_A}{1+b_AP_A+b_BP_B}$$
$$\theta_B = \frac{b_BP_B}{1+b_AP_A+b_BP_B}$$

that is,

$$\theta_i = \frac{b_iP_i}{1+\sum_i b_iP_i}$$

Since

$$\theta_i = \frac{V_i}{V_{m,i}}$$

the expression of the Langmuir adsorption isotherm equation for this gas mixture is just Eq. (10.25), and some times expressed as

$$V_i = \frac{V_{m,i} b_i P_i}{1 + \sum_i b_i P_i} \tag{10.26}$$

10.2.3 Freundlich Adsorption Equation

The variation of adsorption with pressure can often be represented (especially at moderately low pressures) by the equation

$$V = KP^{1/n} \tag{10.27}$$

where k and n are constants, n generally being greater than 1. Taking logarithms

$$\lg V = \lg K + \frac{1}{n} \lg P \tag{10.28}$$

that is, a plot of log V versus log P should give a straight line, and

$$\text{Slope} = \frac{1}{n}$$

$$\text{Intercept} = \lg K$$

Alternatively one can make a linear fitting by Excel to obtain the slope, the intercept, as well as the multiple correlation coefficient R^2, which can be used to determine the degree of fitting, thus judge if the data obtained in the experiment conform with the Freundlich equation.

This adsorption equation was originally proposed on a purely empirical basis. It can be derived theoretically, however, for an adsorption model in which the magnitude of the heat of adsorption varies exponentially with surface coverage. The Freundlich equation is, in effect, the summation of a distribution of Langmuir equations; however, the volume of gas adsorbed is not depicted as approaching a limiting value as in a single Langmuir equation.

10.2.4 Temkin Adsorption Equation

Since the surface is not uniform everywhere, as described previously, the Langmuir equation in some cases does not conform very well with the data obtained in the experiment. To solve this problem, suppose the heat of adsorption linearly decreases with the increase in θ, that is,

$$q = q_0 (1 - \alpha\theta) \tag{10.29}$$

where q_0 is the initial heat of adsorption, that is, the heat of adsorption when $\theta = 0$ and α constant. At equilibrium, Eq. (10.12) derived previously shows

$$a_0(1-\theta)\mu = v_1\theta \tag{10.12}$$

Thus

$$\frac{\theta}{1-\theta} = \frac{a_0\mu}{v_1} \tag{10.30}$$

Substituting Eqs. (10.8) and (10.9) into Eq. (10.30) gives

$$\frac{\theta}{1-\theta} = \frac{a_0}{k_0 e^{-q/K_B T}(2\pi m K_B T)^{1/2}} \cdot P \tag{10.31}$$

Again substituting Eq. (10.29) into Eq. (10.31) gives

$$\frac{\theta}{1-\theta} = \frac{a_0 e^{q_0/K_B T}}{k_0\, e^{q_0\alpha\theta/K_B T}(2\pi m K_B T)^{1/2}} \cdot P$$

Let

$$B_0 = \frac{a_0 e^{q_0/K_B T}}{k_0(2\pi m K_B T)^{1/2}}$$

then

$$\frac{\theta}{1-\theta} = \frac{B_0}{e^{q_0\alpha\theta/K_B T}} \cdot P$$

Taking logarithms,

$$\ln P = -\ln B_0 + \frac{q_0\alpha\theta}{K_B T} + \ln\frac{\theta}{1-\theta} \tag{10.32}$$

Commonly, $q_0\alpha \gg K_B T$ for chemical adsorption, and when θ is in the range of 0.2~0.8, the variation of $\ln[\theta/(1-\theta)]$ with θ is little and approximately equals zero; therefore, the value of ln P mainly depends on $q_0\alpha\theta/K_B T$, thus Eq. (10.32) is simplified as

$$\theta = \frac{K_B T}{q_0\alpha}\ln B_0 P \tag{10.33}$$

which is the Temkin adsorption equation. If Eq. (10.33) is applicable, the plot of ln P versus θ should be a straight line; obviously, when θ approaches zero or 1, it will be not applicable.

10.2.5 BET Equation for Multimolecular Layer Adsorption

Because the force acting in physical adsorption is similar to that operating in liquefaction (ie, van der Waals force), physical adsorption is not limited to a monomolecular layer but can continue until a multimolecular layer of liquid covers the adsorbent surface. The theory of Brunauer, Emmett, and Teller is an extension of the Langmuir treatment to allow for multilayer adsorption at nonporous solid surfaces. The BET equation is derived by balancing the rates of evaporation and condensation for the various adsorbed molecular layers and is based on the simplifying assumption that a characteristic heat E_1 applies to the first monolayer, while the liquefaction heat E_L of the vapor in question applies to adsorption in the second and subsequent molecular layers as follows.

Let the area of surface without adsorption be s_0, the area of surface covered by adsorbed monomolecular layer be s_1, the area of surface covered by two adsorbed molecular layers be s_2, …, and the area of surface covered by i adsorbed molecular layers be s_i.

At equilibrium the area of bare surface is constant, therefore, the rate of condensation to form the first adsorbed molecular layer must be equal to the rate of evaporation to cause the desorption of the first adsorbed molecular layer.

$$a_1 P S_0 = b_1 S_1 \mathrm{e}^{-E_1/RT}$$

where the left side is the rate for forming the first adsorbed molecular layer, in which a_1 is adsorption rate constant and P gas pressure; the right side is the rate of desorption of the first adsorbed molecular layer, in which b_1 is the desorption rate constant of the first adsorbed layer, E_1 solid–gas adsorption heat, and $\mathrm{e}^{-E_1/RT}$ the ratio of molecules of energy above E_1 in the first adsorbed layer according to the Boltzmann distribution. Therefore,

$$S_1 = \left(\frac{a_1}{b_1}\right) P \mathrm{e}^{E_1/RT} S_0 \tag{10.34}$$

Again at equilibrium the area covered by the one adsorbed molecular layer is constant; therefore, the rate of condensation to form the second adsorbed molecular layer must be equal to the rate of evaporation to cause the desorption of the second adsorbed molecular layer.

$$a_2 P S_1 = b_2 S_2 \mathrm{e}^{-E_2/RT}$$

where the left side is the rate for forming the second adsorbed molecular layer, in which a_2 is the adsorption rate constant and P the gas pressure; the right side is the rate of desorption of the second adsorbed molecular layer, in which b_2 is the desorption rate constant of the second adsorbed layer; E_2 the liquid–liquid adsorption heat, that is, the liquefaction heat; and $\mathrm{e}^{-E_2/RT}$ the ratio of molecules of energy above E_2 in the second adsorbed layer according to the Boltzmann distribution. Therefore,

$$S_2 = \left(\frac{a_2}{b_2}\right) P \mathrm{e}^{E_2/RT} S_1 \tag{10.35}$$

Similarly,

$$S_3 = \left(\frac{a_3}{b_3}\right) Pe^{E_3/RT} S_2 \tag{10.36}$$

...

$$S_i = \left(\frac{a_i}{b_i}\right) Pe^{E_i/RT} S_{i-1} \tag{10.37}$$

where E_L is the liquefaction heat, and $E_L = E_2 = E_3 = \cdots = E_i$ (not equal to E_1). Then the total surface of adsorbents can be expressed as

$$A = \sum_{i=0}^{\infty} S_i \tag{10.38}$$

Thus the amount of adsorption is given by

$$V = V_0 \sum_{i=0}^{\infty} iS_i \tag{10.39}$$

where V_0 is the saturated adsorption capacity per unit area. Dividing Eq. (10.39) by Eq. (10.38) gives

$$\frac{V}{AV_0} = \frac{\sum_{i=0}^{\infty} iS_i}{\sum_{i=0}^{\infty} S_i} \tag{10.40}$$

that is,

$$\frac{V}{V_m} = \frac{\sum_{i=0}^{\infty} iS_i}{\sum_{i=0}^{\infty} S_i} \tag{10.41}$$

where V_m equals the volume of gas that would be adsorbed if a monolayer were formed. Obviously, we need to evaluate S_i. First suppose

$$\frac{b_2}{a_2} = \frac{b_3}{a_3} = \cdots = \frac{b_i}{a_i} = g \tag{10.42}$$

$$y = \left(\frac{a_1}{b_1}\right) Pe^{E_1/RT} \tag{10.43}$$

$$x = \left(\frac{a_i}{b_i}\right) P e^{E_L/RT} = \left(\frac{P}{g}\right) e^{E_L/RT} \tag{10.44}$$

$$c = y \cdot x^{-1} = \left(\frac{a_1 g}{b_1}\right) e^{(E_1 - E_L)/RT} \tag{10.45}$$

Then Eqs. (10.34)~(10.37) are simplified as

$$\begin{aligned} S_1 &= yS_0 \\ S_2 &= xS_1 \\ S_3 &= xS_2 = x^2 S_1 \\ &\cdots \\ S_i &= x^{i-1} S_1 = yx^{i-1} S_0 = cx^i S_0 \end{aligned} \tag{10.46}$$

Substituting Eq. (10.46) into Eq. (10.41) and including the bare surface area in the total surface area gives

$$\frac{V}{V_m} = \frac{cS_0 \sum_{i=0}^{\infty} ix^i}{S_0\left(1 + c\sum_{i=1}^{\infty} x^i\right)} \tag{10.47}$$

Since

$$\sum_{i=1}^{\infty} x^i = \frac{x}{1-x} \text{(see Appendix C: (3) Series)} \tag{10.48}$$

and

$$\sum_{i=1}^{\infty} ix^i = x\frac{d}{dx}\sum_{i=1}^{\infty} x^i = \frac{x}{(1-x^2)} \tag{10.49}$$

Combining Eqs. (10.47), (10.48), and (10.49) gives

$$\frac{V}{V_m} = \frac{cx}{(1-x)(1-x+cx)} \tag{10.50}$$

Now let us solve x as follows. Suppose P approaches saturated vapor pressure P_0, V will approach ∞; also it can be seen from Eq. (10.50) when x approaches 1, V will also approach ∞. This implies $x = 1$ when $P = P_0$; therefore, from Eq. (10.44) we have

$$x = \left(\frac{a_i}{b_i}\right) P_0 e^{E_L/RT} = \left(\frac{P_0}{g}\right) e^{E_L/RT} = 1$$

Since dividing x by 1 does not change its value

$$x = \frac{\left(\frac{P}{g}\right) e^{E_L/RT}}{\left(\frac{p_0}{g}\right) e^{E_L/RT}} = \frac{P}{P_0} \tag{10.51}$$

Substituting Eq. (10.51) into Eq. (10.50) gives

$$V = \frac{V_m cP}{(P_0 - P)\left[1 + (c-1)\frac{P}{P_0}\right]} \tag{10.52}$$

which is referred to as the BET equation for multimolecular layer adsorption, where V is the amount of adsorption when gas pressure equals P; V_m the amount of adsorption at monolayer coverage; and c the adsorption coefficient, which is the function of heat of adsorption as shown by Eq. (10.45), therefore, represents the strength of gas adsorption on solid. The value of c has important influence on the shapes of adsorption isotherms, as shown in Fig. 10.5. If the value of c is great, the adsorption of gas molecules on solid surfaces will be much stronger than on the monolayer formed; therefore, the adsorption of gas on solid surfaces takes place preferentially, and the monolayer coverage can be completed when $P \ll P_0$ before the second layer of adsorption starts. After the monolayer coverage is completed, another layer of adsorption will start, as shown by the curve for $c = 100$ in Fig. 10.5. In such a case the Eq. (10.52) can be simplified as the Langmuir monomolecular adsorption equation.

$$V = \frac{V_m cP}{(P_0 - P)\left[1 + (c-1)\frac{P}{P_0}\right]} \approx V_m \frac{cP}{p_0\left(1 + c \cdot \frac{P}{P_0}\right)} = \frac{V_m cP}{P_0 + cP} = \frac{V_m bP}{1 + bP}$$

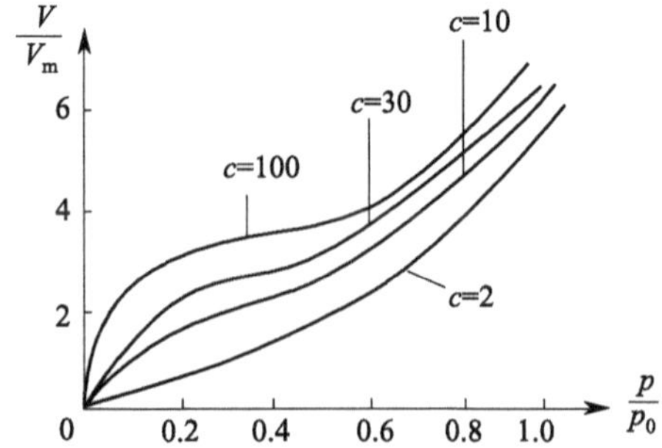

FIGURE 10.5 **Effect of adsorption coefficient on the shapes of adsorption isotherms.**

where $b = \frac{c}{P_0}$. If the value of c is small, another layer of adsorption will start nearly as the same time as the adsorption of gas on a solid surface is taking place; thus, the platform of the Langmuir curve will disappear, as shown by the curves for $c = 30$, 20, and 2.

The BET equation for multimolecular layer adsorption represents curve I, curve II, and curve III in Fig. 10.4. Rearranging Eq. (10.52) gives the linear equation

$$\frac{P}{V(P_0 - P)} = \frac{1 + (c-1)\frac{P}{P_0}}{V_m c} = \frac{1}{V_m c} + \frac{c-1}{V_m c} \cdot \frac{P}{P_0} \tag{10.53}$$

This form suggests that the plot of $\frac{P}{V(P_0 - P)}$ versus $\frac{P}{P_0}$ should be a straight line, and from the slope and intercept of the line, we obtain

$$\text{Slope} = \frac{c-1}{V_m c} \tag{10.54}$$

$$\text{Intercept} = \frac{1}{V_m c} \tag{10.55}$$

Combining Eqs. (10.54) and (10.55) gives the amount of adsorption at monolayer coverage V_m and adsorption coefficient c. Alternatively we can make a linear fitting by using Excel to obtain the slope and the intercept, as well as the multiple correlation coefficient R^2, which can be used to determine the degree of fitting, and then judge if the data obtained in the experiment conforms with the BET equation. In addition, if you multiply the V_m by the area of cross-section of one molecule of adsorbed gas (for example N_2), we will obtain the specific area of solid.

It is noted that there are some deviations between BET equation and practical data. The reason is that some assumptions which are not in accord with practices were made in the derivation of the equation; for example, the heat of adsorption is the same everywhere on the surface and there are no interactions between the adjacent molecules adsorbed at the surface. Especially the latter is in conflict with the multiplayer adsorption. If there were no interactions between the adjacent molecules adsorbed at the surface, the adsorptions of the second layer and the layers above the second layer would not be able to take place.

Example 10.3 BET Equation for Multimolecular Layer Adsorption

Making a linear fitting for the BET equation and the adsorption data of N_2 gas on a solid adsorbent at the temperature of 77 K gives the slope of 1.50 g cm^{-3} and the intercept of 0.005 g cm^{-3}, respectively. Calculate the monolayer adsorption capacity V_m and the specific area S, and the heat of adsorption for the first layer adsorption respectively. (The heat of liquefaction of N_2 gas is known as 5.43 kJ mol^{-1} and the area of cross-section of one molecule of N_2 represented by δ_m is known as 0.162×10^{-18} m^2.)

Solution

According to Eq. (10.52)

$$\frac{P}{V(P_0 - P)} = \frac{1}{V_m c} + \frac{c-1}{V_m c} \cdot \frac{P}{P_0}$$

Plotting the straight line of $\frac{P}{V(P_0 - P)}$ versus $\frac{P}{P_0}$ gives

$$\text{Slope} = \frac{c-1}{V_m c} \tag{10.54}$$

$$\text{Intercept} = \frac{1}{V_m c} \tag{10.55}$$

Combining Eqs. (10.54) and (10.55) gives

$$V_m = \frac{1}{\text{intercept} + \text{slope}} = \frac{1}{0.005 + 1.50} = 0.664\ (\text{cm}^3\ \text{g}^{-1})$$

$$c = \left(\frac{\text{slope}}{\text{intercept}}\right) + 1 = 301$$

thus

$$S = \frac{N_A \cdot V_m \cdot \sigma_m}{22{,}400} = \frac{6.02 \times 10^{23} \times 0.664 \times 0.162 \times 10^{-18}}{22{,}400} = 2.89\ (\text{m}^2\ \text{g}^{-1})$$

Since

$$c \approx \exp[(E_1 - E_L)/RT]$$

$$E_1 = RT \ln c + E_L = 9092\ (\text{J mol}^{-1})$$

10.2.6 Polanyi Adsorption Potential Theory and D–R Equation

The theory of Polanyi adsorption potential does not concern the physical image of adsorption; for example, the surface is uniform or not uniform, and the adsorption is monolayer or multilayers. The main viewpoints are introduced as follows.

1. There is an adsorption force field in a space region near the solid surface. Once a molecule falls into this region, it will be adsorbed. This space region is called adsorption space.

2. There is an adsorption potential at any location in the adsorption space. This adsorption potential is defined as the isothermal reversible nonvolume work required to adsorb 1 mole of a gas from a location outside the adsorption space to a location inside the adsorption space, and commonly expressed by ε_x

$$\varepsilon_x = -\Delta G = \int_p^{p_0} V\,dp = RT \ln \frac{p_0}{p} \tag{10.56}$$

where x represents the distance from the solid surface, P the pressure of gas at equilibrium, P_0 the saturated vapor pressure, and V the mole volume of the adsorbate. The surface formed by the points of the same adsorption potential is referred to as the isopotential surface. The volume between the isopotential surface and the solid surface is referred to as adsorption volume, that is, the amount of adsorption. Suppose the adsorption volume between the solid surface and the isopotential surface at distance x is V_x, then we have

$$V_x = \frac{aM}{\rho} = a\bar{V} \tag{10.57}$$

where a is the experimentally measured amount of adsorption (mol), ρ the density of liquid adsorbate at the experiment temperature, and $\bar{V}$ the mole volume of liquid adsorbate. Eqs. (10.56) and (10.57) are basic equations of Polanyi adsorption potential theory.

3. It is assumed that adsorption potential and its distribution curve in adsorption space is independent of temperature; that is, the relationships between ε_x and V_x at different temperatures are the same. Therefore the curve for the relationship between ε_x and V_x is referred to as the characteristic curve of the adsorbate–adsorbent system. Fig. 10.6 illustrates the characteristic curve of CO_2 adsorption at carbon's surface at five different temperatures. Essentially, the characteristic

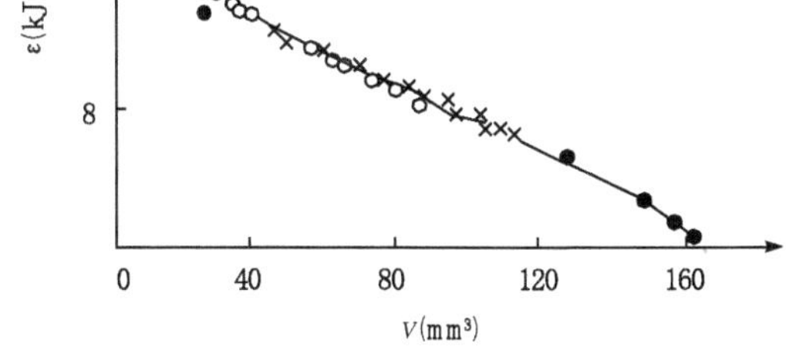

FIGURE 10.6 Characteristic curve of CO_2 adsorption at carbon surface.

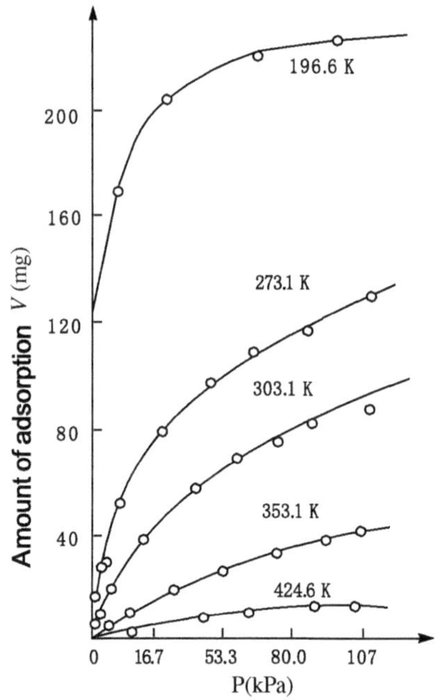

FIGURE 10.7 **Adsorption isotherm of CO_2 at carbon surface at different temperatures.**

curve is the adsorption isotherm because ε_x is the function of pressure P, and V_x is related to amount of adsorption a, as shown by Eqs. (10.56) and (10.57).

Since the characteristic curve is independent of temperature, we can take a characteristic curve transformed from an adsorption isotherm at a temperature and calculate the pressure and amount of adsorption at another temperature, and then obtain the adsorption isotherm at another temperature. Fig. 10.7 illustrates the adsorption isotherm of CO_2 at carbon's surface at different temperatures which are obtained from experimentally measured adsorption isotherms at 273.1 K.

It was proven that the ratios of adsorption potentials for equal adsorption volumes at the characteristic curves of different adsorbates for the same adsorbent are constant. This ratio is termed coefficient of affinity and represented by β, as shown in Eq. (10.57) and Fig. 10.8.

$$\frac{\varepsilon_1}{\varepsilon_2} = \frac{\varepsilon_3}{\varepsilon_4} = \cdots = \beta \tag{10.58}$$

The coefficient of affinity can be calculated from many kinds of physicochemical constants of the adsorbate. For the same adsorbent, if one kind of adsorbate is taken as the reference substance, the coefficient of affinity of another adsorbate can be worked out. Table 10.2 lists the coefficients of affinity of many kinds of gases on activated carbon, which are obtained when that of benzene is taken as 1.

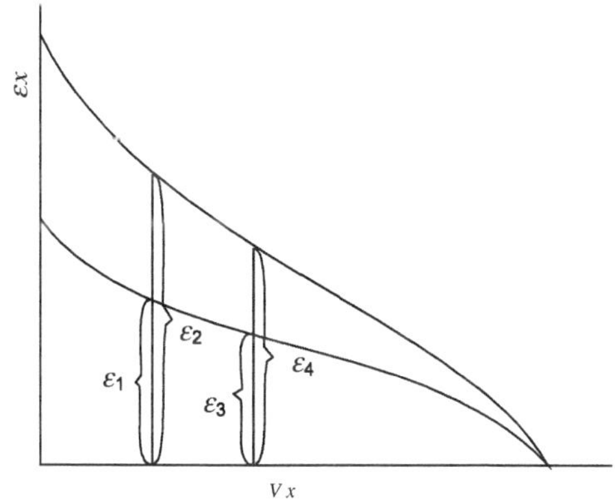

FIGURE 10.8 Determination of the coefficient of affinity.

Dubinin et al. classified activated carbon into two types. The first one mainly contains micropores; the second one contains larger pores. Based on the careful analysis, Dubinin et al. gave the expression of the characteristic curve for the first type of activated carbon

$$V = V_0 \exp(-k\varepsilon^2) \tag{10.59}$$

$$\ln V = \ln V_0 - k\varepsilon^2 \tag{10.60}$$

where V is the adsorption volume when adsorption potential is ε, k the constant which is related to the size of micropores, and V_0 the volume of micropores. If the specific area of mesopores is smaller than 50 $m^2\ g^{-1}$, V_0 will be considered to be equal to the volume of all pores in the activated carbon. Substituting Eqs. (10.56) and (10.57) into Eq. (10.59) gives

$$a = \frac{V_0}{\overline{V}} \exp\left[-k\left(RT \ln \frac{p_0}{p}\right)^2\right] \tag{10.61}$$

Taking the natural logarithm of it gives

$$\ln a = \ln \frac{V_0}{\overline{V}} - kR^2T^2\left(\ln \frac{p_0}{p}\right)^2 \tag{10.62}$$

The plot of ln a versus $\left(\ln \frac{p_0}{p}\right)^2$ should be a straight line, and from the intercept and slope of the line, we can obtain the volume of micropores V_0 and the constant k. Taking the logarithm of Eq. (10.61) and combining the constants gives

$$\lg a = C - D\left(\lg \frac{p_0}{p}\right)^2 \tag{10.63}$$

TABLE 10.2 Coefficients of Affinity of Different Gases on Activated Carbon

Gas	Benzene	C_5H_{12}	C_6H_{12}	C_7H_{16}	CH_3Cl	$CHCl_3$	CCl_4	CH_3OH	C_2H_5OH	$HCOOH$	CH_3COOH	$(C_2H_5)_2O$	CS_2	NH_3
β	1	1.12	1.04	1.50	0.56	0.88	1.07	0.40	0.61	0.60	0.97	1.09	0.70	0.28

Benzene is taken as the reference substance.

where

$$C = \lg\left(\frac{V_0}{\overline{V}}\right)$$

$$D = KkR^2T^2$$

where K is a constant. Eq. (10.59), Eq. (10.62), and Eq. (10.63) are all referred to as Dubinin–Radushkevich equations.

10.2.7 Capillary Condensation and Adsorption Hysteresis

1. Capillary condensation
   It has previously been shown (Section 8.3) that the vapor pressure over a convex liquid surface is greater than that over the corresponding flat surface. A liquid which wets the wall of a capillary will have a concave liquid–gas interface and, therefore, a lower vapor pressure in the capillary than in the bulk phase. This lower vapor pressure is given by the Kelvin equation, written in the form

$$RT\ln\frac{P_r}{P_0} = -\frac{2\gamma V\cos\theta}{R} \tag{10.64}$$

   where R is the radius of the capillary, and θ the contact angle between the liquid and the capillary wall. The minus added to the right side of the equation is due to the negative value of the radius of curvature for concave surfaces.

   Condensation can, therefore, take place in narrow capillaries at pressures which are lower than the normal saturation vapor pressure, as shown in Fig. 10.9. Zsigmondy suggested that this phenomenon might also apply to porous solid.

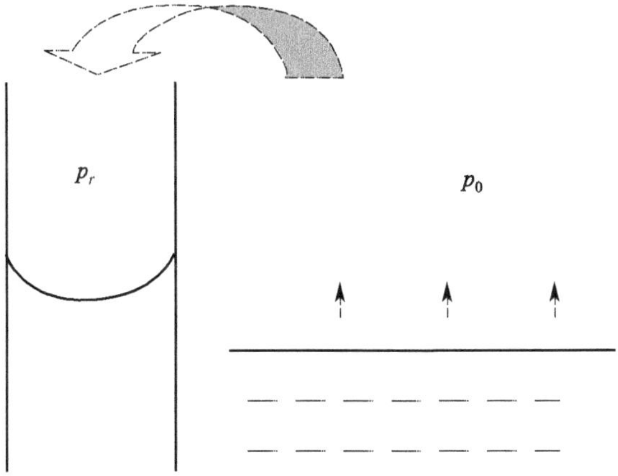

FIGURE 10.9 Capillary condensation.

It can be seen from Eq. (10.64) that if $P_r < P_0$, the vapor outside the capillary will enter into it, causing the capillary condensation. The larger the radius of the capillary, the greater the P_r. When the radius is increased so that $P_0 = P_r$, the capillary condensation will stop, and only normal adsorption will act.

Example 10.4 Capillary Condensation

Calculate the radius of the capillary at which the capillary condensation of nitrogen gas takes place. Suppose the temperature is 77 K, $P_r/P_0 = 0.75$, the surface tension of liquid nitrogen is 8.85 mN m^{-1}, and mole volume of liquid nitrogen is 34.7 cm^3.

Solution

Suppose the contact angle is 0 degrees; according to the Kelvin equation

$$RT\ln\frac{P_r}{P_0} = -\frac{2\gamma V\cos\theta}{R}$$

$$8.314\times 77\ln 0.75 = -\frac{2\times 8.85\times 10^{-3}\times 34.7\times 10^{-6}}{R}$$

$$R = 3.34\times 10^{-9}\ \text{(m)} = 3.34\ \text{(nm)}$$

2. Adsorption hysteresis
   The capillary condensation theory provides a satisfactory explanation of the phenomenon of adsorption hysteresis, which is frequently observed for porous solids. Adsorption hysteresis is a term which is used when the desorption isotherm curve does not coincide with the adsorption isotherm curve, as shown in Fig. 10.10.

   A possible explanation of this phenomenon is given in terms of contact angle hysteresis. The contact angle during adsorption, when liquid is advancing over a dry surface, is generally greater than the contact angle during desorption, when liquid is receding from a wet surface. Generally, the former is termed advancing contact angle, and the latter is termed receding

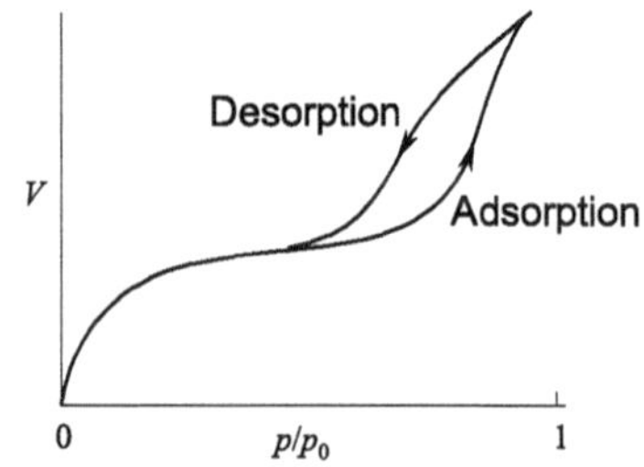

FIGURE 10.10 Adsorption hysteresis.

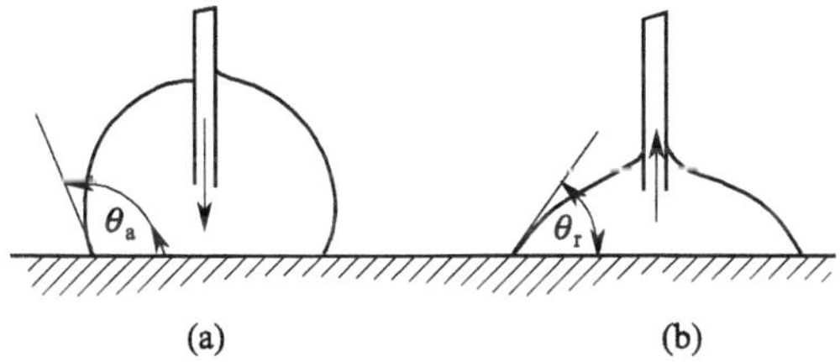

FIGURE 10.11 (a) Advancing contact angle and (b) receding contact angle.

contact angle. Fig. 10.11 schematically shows the advancing contact angle and the receding contact angle.

From the Kelvin equation, it is evident that the pressure below which liquid vaporizes from capillaries will, under these circumstances, be lower than the pressure required for capillary condensation.

10.3 ADSORPTION FROM SOLUTION

10.3.1 Amount of Adsorption

The theoretical treatment of adsorption of materials from solution to solid surface is generally more complicated than that of gas adsorption, since adsorption from solution always involves competition between solute(s) and solvent or between the components of a liquid mixture for the adsorption sites. Consider, for example, a binary liquid mixture in contact with a solid. Zero adsorption refers to uniform mixture composition right up to the solid surface, even though both components are in fact present at the solid surface. If the proportion of one of the components at the surface is greater than its proportion in bulk, then that component is positively adsorbed and, consequently, the other component is negatively adsorbed. Apparent, rather than true, adsorption isotherms are, therefore, calculated from changes in solution concentration as shown in Eq. (10.65).

$$q = \frac{x}{m} = \frac{V(c_0 - c)}{m} \tag{10.65}$$

where q is the amount of adsorption, x the mass of materials adsorbed, m the mass of adsorbent, V the volume of solution, c_0 the concentration of the component before adsorption, and c the concentration of the component after adsorption.

Thermodynamics of adsorption have been discussed in Section 10.1.2. For the adsorption from solution we only need to substitute concentration c for the pressure P and mass q for the volume V at the standard state in those equations.

1. Adsorption in the total concentration range
   Examples of apparent adsorption isotherms for binary liquid mixtures in the total concentration range are given in Fig. 10.12.

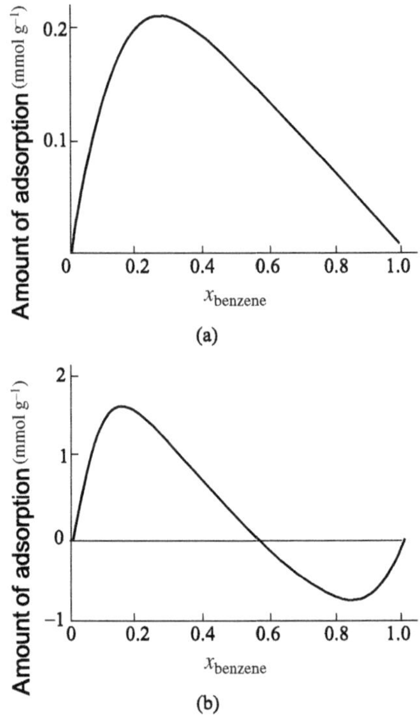

FIGURE 10.12 **Isotherms for the adsorption of benzene from solution.** (a) U-type and (b) S-type.

In Fig. 10.12 x is the mole fraction of solute at equilibrium. It shows the U-type isotherm and the S-type isotherm respectively in the range of x from 0 to 1.

2. Adsorption from dilute solution

The isotherms of adsorption from dilute solution are very similar to those of gas adsorption. The gas adsorption isotherm equations, including Langmuir, Temkin, Freundlich, BET, and D–R equations, can be used for the adsorption from solution. But we need to substitute concentration c for the pressure P and mass q for the volume V at the standard state in those equations. The BET equation applies to nonporous adsorbents or porous adsorbents of pore size much larger than the size of the adsorbate molecules. For the Polanyi adsorption potential theory and D–R equation, Eq. (10.60) is expressed as

$$\ln q_{\mathrm{e}} = \ln q_{\mathrm{m}} - \beta\varepsilon^2 \tag{10.66}$$

where

$$\varepsilon = RT\ln\left(1+\frac{1}{c_e}\right) \tag{10.67}$$

$$E_a = \sqrt{2\beta} \tag{10.68}$$

Where c_e is the concentration of solute in solution at equilibrium (mg L^{-1}), q_e the amount of adsorption at equilibrium (mg g^{-1}), q_m the adsorption capacity (mg g^{-1}), β the Dubinin–Radukevich constant (mol^2 kJ^{-2}), ε the Polanyi adsorption potential (J mol^{-1}), and E_a the average adsorption energy (kJ mol^{-1}) which reflects the adsorption mechanism. If $E_a < 8$ kJ mol^{-1}, the main adsorption mechanism might be physical; if $8 < E_a < 16$ kJ mol^{-1}, the main adsorption mechanism might be ion exchange; and if $E_a > 16$ kJ mol^{-1}, the main adsorption mechanism might be the strong chemical adsorption.

Example 10.5 Adsorption from Solution

The Langmuir equation can be used to describe the adsorption behavior of suspended substances. If the concentration of solute at adsorption equilibrium is 3.0×10^{-3} mol dm^{-3}, the decrease in the concentration of solute due to the adsorption by 1 g of suspended substances will be 0.5×10^{-3} mol dm^{-3}; if the concentration of solute at adsorption equilibrium is 1.0×10^{-3} mol dm^{-3}, the decrease in the concentration of solute due to adsorption by 1 g of suspended substances will be 0.25×10^{-3} mol dm^{-3}. Calculate the maximum amount of adsorption per gram of suspended substance.

Solution

Langmuir equation can be expressed as

$$q = \frac{q_m bc}{1 + bc}$$

Therefore

$$0.5 \times 10^{-3} = \frac{q_m b \times 3.0 \times 10^{-3}}{1 + b \times 3.0 \times 10^{-3}}$$

$$0.25 \times 10^{-3} = \frac{q_m b \times 1.0 \times 10^{-3}}{1 + b \times 1.0 \times 10^{-3}}$$

Combining and solving these two equations gives the maximum amount of adsorption per gram of suspended substance

$$q_m = 1.0 \times 10^{-3} \left(\text{mol dm}^{-3}\ \text{g}^{-1}\right)$$

Example 10.6 Adsorption from Solution

The adsorption of a solute from a solution onto carbon obeys the Langmuir equation. The adsorption capacity is known as $q_m = 4.2$ mmol g^{-1}; adsorption coefficient $b = 2.8$ cm^3 $mmol^{-1}$. Add 5 g of carbon into 200 cm^3 of solution ($c = 0.2$ mol dm^{-3}) and evaluate the concentration of the solution at equilibrium.

Solution

The Langmuir equation can be expressed as

$$q = \frac{q_m bc}{1+bc} = \frac{V(c_0 - c)}{m},$$

therefore

$$\frac{4.2\times10^{-3}\times2.8c}{1+2.8c} = \frac{0.2(0.2-c)}{5}$$

$$c = 0.1665\ (\text{mol dm}^{-3})$$

10.3.2 Adsorption Kinetics

Usually, the surface adsorption is divided into the following elementary steps: (1) The molecules or ions of the adsorbate diffuse from the solution onto the surface of the adsorbent; (2) for multiporous adsorbent, the molecules or ions of the adsorbate diffuse in pores to the adsorption site; and (3) the molecules or ions of the adsorbate combine with the adsorption sites of the surface.

Generally, there is a stationary liquid film near the surface of the adsorbent. The molecules or ions of the adsorbate must pass through this film to reach the surface. Since the resistance of liquid film for diffusion is greater than that of the solution, the diffusion in liquid film might be the rate determine step. But strong disturbances can reduce the resistance of the film for diffusion; as a result, the diffusion in the liquid film would not play the decisive effect any longer.

The adsorption rate of nonporous adsorbent is very great. Generally, it completes 90~95% adsorption within 1~2 s and reaches adsorption equilibrium within 10~20 s; whereas the adsorption rate of porous adsorbents is very low. It needs a very long time to reach equilibrium, which indicates that the diffusion of adsorbate in pores has a great effect on adsorption rate. During the diffusion process, the molecules or ions of the adsorbate might be temporarily adsorbed by the adsorption site on the wall of the pores, thus delaying the adsorption, reducing the effective cross-section of pores, and hindering the diffusion of other molecules or ions. When the volume of molecules or ions of the adsorbate is large, the diffusion in pores often becomes the rate determine step.

The diameter of adsorbent particles has a great effect on adsorption rate. The smaller the particle, the shorter the time needed to reach adsorption equilibrium. For a nonporous adsorbent, the adsorption rate is inversely proportional to the diameter of particles. For a multiporous adsorbent, if the diffusion in surface film is rate determine step, the rate of adsorption also will be inversely proportional to the diameter of particles, as in the situation of nonporous adsorbents, whereas, if the diffusion in pores is the rate determine step, the rate of adsorption will be inversely proportional to the high order power of diameter of particles. The effects of pore size on the rate of adsorption and the amount of adsorption are contradictory. Although small pore size makes the rate of adsorption slower, it leads to a higher amount of adsorption. Although a higher portion of pores of large size is favorable for increasing the rate of adsorption, it leads to a lower amount of adsorption.

The size of molecules or ions of the adsorbate also has an important effect on the rate of adsorption. For example, when activated carbon is used to adsorb alkyl benzene sulfonic acid, the inner diffusion plays a main effect. It will be discovered that the more the carbon atoms of alkyls, the lower the rate of adsorption.

10.3.3 Pseudofirst Order Model

The Lagergren pseudofirst order equation is the earliest model which describes the rate of adsorption at gas–liquid interface. The basic principle is that the rate of adsorption is proportional to the number of bare adsorption site on the surface, that is,

$$\frac{dq_t}{dt} = k_1(q_e - q_t) \tag{10.69}$$

where q_t is the amount of adsorption (mmol g^{-1}) at contact time t, q_e the amount of adsorption (mmol g^{-1}) at equilibrium, and k_1 the pseudofirst order rate constant (min^{-1}). Integrating it from $t = 0$ to $t = t$ and $q_t = 0$ to $q_t = q_t$, respectively gives

$$q_t = q_e(1 - e^{-k_1 t}) \tag{10.70}$$

The linear equation is given by

$$\ln(q_e - q_t) = \ln q_e - k_1 t \tag{10.71}$$

The pseudofirst order model cannot be applied to describing the ideal monolayer adsorption in which the rapid adsorption dominates, whereas the slow adsorption is weak. For most situations, it does not apply to the total adsorption process, but only applies to the initial stage of 20~30 min of adsorption.

10.3.4 Pseudosecond Order Model

In the pseudosecond order model, the rate determine step is considered to be the chemical reaction or chemisorption. The basic principle is that the rate of adsorption is proportional to the second power of the number of bare adsorption site on the surface, that is,

$$\frac{dq_t}{dt} = k_2(q_e - q_t)^2 \tag{10.72}$$

where q_t is the amount of adsorption (mmol g^{-1}) at contact time t, q_e the amount of adsorption (mmol g^{-1}) at equilibrium, and k_2 the pseudosecond order rate constant (g mmol^{-1} min^{-1}). Integrating it from $t = 0$ to $t = t$ and $q_t = 0$ to $q_t = q_t$, respectively gives

$$q_t = \frac{q_e^2 k_2 t}{1 + q_e k_2 t} \tag{10.73}$$

This model contains all steps including the diffusion in liquid film, the adsorption at the outer surface, the intraparticle diffusion, and the adsorption at the inner surface. The equation describes the adsorption process, which consists of both rapid adsorption and slow adsorption very well. The linear equation is given by

$$\frac{t}{q_t} = \frac{1}{k_2 q_e^2} + \frac{t}{q_e} \tag{10.74}$$

10.3.5 Intraparticle Diffusion Model

Since both the pseudofirst order equation and pseudosecond order equation cannot explain the adsorption progress of the adsorbate intraparticle of the adsorbent, the intraparticle diffusion model was proposed to fit its data. The linear equation is given by

$$q_t = kt^{1/2} + c \tag{10.75}$$

where q_t is the amount of adsorption (mmol g^{-1}) at contact time t, c is a constant which reflects the thickness of liquid film, the greater the value of c, the greater the effect of the diffusion in liquid film, and k (mg g^{-1} min^{-1}) is the apparent rate constant which is related to the intraparticle diffusion coefficient as follows

$$k = \frac{6q_e}{R}\sqrt{\frac{D}{\pi}}$$

where q_e is the amount of adsorption (mmol g^{-1}) at equilibrium, R the radius of the particle, and D the diffusion coefficient of the intraparticle diffusion. Generally, the total adsorption process involves many kinds of kinetics mechanisms, including the diffusion in liquid film, intraparticle diffusion, and surface diffusion. For many adsorptions, the diffusion in liquid film is the rate determine step. The fitting line which does not pass through the original point implies that intraparticle diffusion is not the only mechanism, and the rate of adsorption is also influenced by the diffusion outside particle. If intraparticle diffusion is the only mechanism, it will take the form

$$q_t = kt^{1/2} \tag{10.76}$$

10.3.6 Elovich Model

In the Elovich model, it is believed that the adsorption rate decreases exponentially with the increase in the amount of adsorption at the surface; the expression is given by

$$\frac{dq_t}{dt} = \alpha \exp(-\beta q_t) \tag{10.77}$$

where q_t is the amount of adsorption (mg g^{-1}) at contact time t, α the initial adsorption-rate constant (mg g^{-1} min^{-1}), β the constant (g mg^{-1}) which is related to activation energy and surface coverage or is termed the desorption-rate constant. Assuming $\alpha\beta >> 1$, when $t = 0$, $q_t = 0$; when $t = t$, $q_t = q_t$, then Eq. (10.77) can be simplified as the linear equation

$$q_t = \left(\frac{1}{\beta}\right)\ln(\alpha\beta) + \left(\frac{1}{\beta}\right)\ln t \tag{10.78}$$

or

$$q_t = a + b\lg t \tag{10.79}$$

where b is a constant which measures the velocity of the decrease in adsorption rate with time. The above equations describe the diffusion mechanisms of the adsorption consisting of both rapid reaction and slow reaction.

As in the case of the adsorption isotherm, we can also make a linear fitting by using Excel to obtain the slope and the intercept, as well as the multiple correlation coefficient R^2, which can be used to determine the degree of fitting, thus, judge if the data obtained in the experiment conforms with the studied equations, and then make a choice of models for the system.

10.4 WETTING OF SOLID SURFACE

Wetting is a common phenomenon in nature. It is often observed in the morning that the water on the surface of leaves of plants appears as dewdrops, which indicates the leaves cannot be wetted by water. When people put their hand into water, the hand will become wet, which indicates the hand can be wet by water. Wetting can be considered the process of substituting a liquid–solid interface for a gas–liquid interface and a gas–solid interface. It is illustrated in Fig. 10.13.

Suppose the wetting shown in Fig. 10.13 takes place spontaneously, and the surface area of every interface is 1 cm^2, then we have

$$\Delta G = \gamma_{s1} - (\gamma_{sA} + \gamma_{1A}) < 0 \tag{10.80}$$

or

$$-\Delta G = \gamma_{sA} + \gamma_{1A} - \gamma_{s1} = W_{s1} \tag{10.81}$$

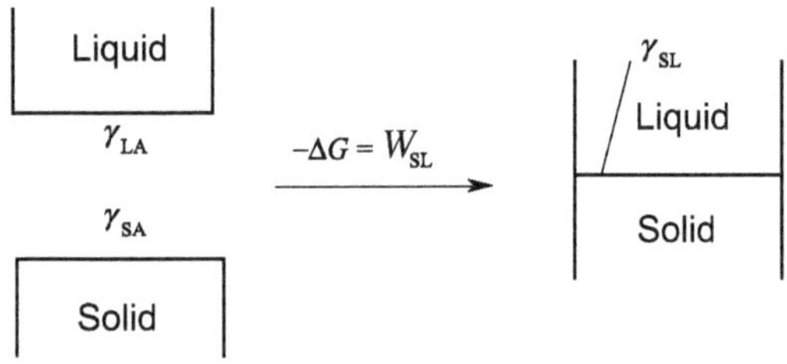

FIGURE 10.13 Essence of wetting.

where W_{sl} is the adhesion work between liquid and solid, which can be used to measure the wetting extent. The greater the adhesion work, the greater the extent of wetting. Because there is not any methods to measure γ_{sA} and γ_{sl} correctly up to now, Eq. (10.81) cannot actually be used to study the wetting extent. But employing the contact angle can solve this problem.

Suppose a drop of liquid is placed on a flat solid surface, or a gas bubble is placed under a flat solid surface in liquid, as shown in Fig. 10.14.

Let us make a tangent line of the air–liquid interface through the junction of three phases, thus we will obtain the contact angle θ, which is the angle between the tangent line and the solid surface in the liquid (see Section 8.4). Then the wetting and nonwetting are relatively defined as:

$$\theta < 90 \text{ degrees: wetting} \quad \theta > 90 \text{ degrees: nonwetting}$$

Consider the forces applied to the junction of three phases, which are shown in Fig. 10.15.

FIGURE 10.14 Contact angle. (a) $\theta < 90$ degrees and (b) $\theta > 90$ degrees.

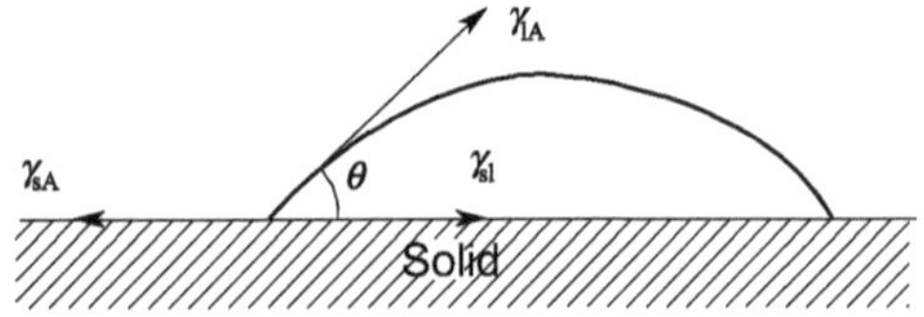

FIGURE 10.15 Force balance at the junction of three phases.

The gas–solid interface tension tends to drag the point O to the left to decrease the area of the gas–solid interface, thus decreasing its surface free energy; similarly, the liquid–solid interface tension tends to drag the point O to the right to decrease the area of the liquid–solid interface, thus decreasing its surface free energy. In addition, the gas–liquid interface tension in the direction of the tangent line of the air–liquid interface tends to drag the point O to the upper right to decrease the area of the gas–liquid interface, thus decreasing its surface free energy. Viewing the surface tensions as forces acting along the perimeter of the drop enable us to write immediately an equation that describes the equilibrium force balance in the horizontal direction:

$$\gamma_{sA} = \gamma_{sl} + \gamma_{lA} \cos\theta \tag{10.82}$$

This result is generally known as Young's equation. Substituting Eq. (10.82) into Eq. (10.81) gives

$$W_{sl} = \gamma_{sl} + \gamma_{lA} \cos\theta + \gamma_{lA} - \gamma_{sl} = \gamma_{lA}(1 + \cos\theta) \tag{10.83}$$

thus we are able to obtain the adhesion work by measuring only the contact angle. It can be seen from Eq. (10.83) that the smaller the contact angle, the greater is the adhesion work, that is, the higher is the wetting extent.

Example 10.7 Contact Angle

It is known that aluminum oxide porcelain needs coating silver on its surface. Can liquid silver wet the surface of aluminum oxide porcelain when it is heated to 1000°C? Suppose γ (Al_2O_3, s) = 1×10^{-3} N m^{-1}, γ (Ag, l) = 0.92×10^{-3} N m^{-1}, and γ (Ag, l/Al_2O_3, s) = 1.77×10^{-3} N m^{-1} at 1000°C.

Solution

According to Eq. (10.82)

$$\cos\theta = \frac{\gamma_{sA} - \gamma_{sl}}{\gamma_{lA}} = \frac{1\times10^{-3} - 1.77\times10^{-3}}{0.92\times10^{-3}} = -0.837$$

$$\theta = 147^\circ > 90^\circ$$

Therefore liquid silver cannot wet the surface of aluminum oxide porcelain.

10.4.1 Measurement of Contact Angle

1. Adam–Jessop method
   The measurement of contact angle is facilitated if a moderately large and uniform solid surface is available. One method is to adjust the angle of a plate immersed in the liquid so that the liquid surface remains perfectly flat right up to the solid surface, as shown in Fig. 10.16; the angle between the liquid surface and the plate is the contact angle, which can be obtained by measuring its vertically opposite angle.

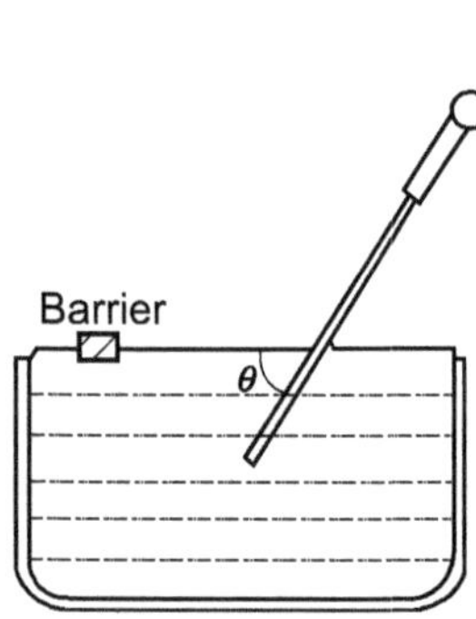

FIGURE 10.16 **Adam–Jessop method for measuring contact angles.**

2. Projection method
   Project the droplet in Fig. 10.15 to a projection screen by monochromatic light, then measure the included angle between the tangent line and the solid surface in the liquid, thus obtaining the contact angle.
3. Bartell–Osterhof method
   The contact angles of solids in a finely divided form are technically important but are difficult to measure. Bartell et al. have developed a method based on displacement pressures for measuring such contact angles. The finely divided solid is packed into a tube and the resulting porous plug is considered to act as a bundle of capillaries of some average radius r. The pressure required to prevent the liquid in question from entering the capillaries of this porous plug is measured, and the general equation for capillary rise is applied

$$P = \Delta P = \frac{2\gamma}{r} = \frac{2\gamma}{\left(\frac{R}{\cos\theta}\right)} = \frac{2\gamma\cos\theta}{R} \tag{10.84}$$

   where r is the radius of curvature of concave surface of the liquid in capillaries and R the equivalent radius of the capillaries. (The relationship between r and R is shown in Fig. 8.6). The equivalent radius of the capillaries is then found from a similar experiment using a liquid, which completely wets the solid ($\theta = 0$), that is,

$$P' = \frac{2\gamma'}{R} \tag{10.85}$$

Then we can calculate the contact angle θ from Eq. (10.84) for any other liquid.

If the pressure described earlier is not applied, the liquid in question will enter into the capillaries of this porous plug until the equilibrium is reached,

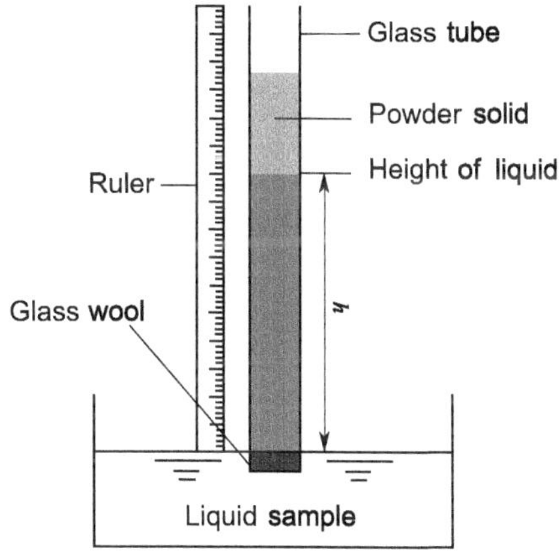

FIGURE 10.17 Method for measuring the contact angle of powder solids.

as shown in Fig. 10.17, measuring the height of capillary rise at equilibrium and equating the static pressure of the liquid column, and the extra pressure will give the contact angle.

Washburn's equation relates the capillary rise height of liquid through a compact vertical bed of particles with small pore radius and contact angle as follows

$$h^2 = (\gamma R \cos\theta/2\eta)t \tag{10.86}$$

where h is the height of liquid penetrating the bed in time t, η the viscosity of the penetrating liquid, γ the surface tension of the penetrating liquid, R the capillary radius, and θ the advancing particle contact angle. Plotting the straight line of h^2 versus time t gives the slope k, which is expressed as

$$k = \gamma R \cos\theta/2\eta \tag{10.87}$$

Thus

$$\theta = \arccos(2k\eta/\gamma R) \tag{10.88}$$

Example 10.8 Measurement of Contact Angle

The pressures required to prevent two liquids from entering the capillaries of a porous plug are measured respectively. The surface tension and the contact angle of the first one are known as 50 mN m^{-1} and 0 degree, respectively. The surface tension of the second one is known as 70 mN m^{-1}. The measurement shows that the pressure required for the first one is two times that of the second one. Estimate the contact angle of the second liquid on the surface of the porous plug.

Solution
The pressure required for the first liquid can be expressed as

$$P_1 = \frac{2\gamma_1 \cos\theta_1}{R_1} = \frac{2\gamma_1 \cos 0^\circ}{R_1}$$

The pressure required for the second liquid can be expressed as

$$P_2 = \frac{2\gamma_2 \cos\theta_2}{R_2}$$

When the equivalent radii of the capillaries are very small, we take $R_1 \approx R_2$. Therefore,

$$\frac{P_1}{P_2} = \frac{\gamma_1}{\gamma_2 \cos\theta_2} = 2$$

$$\cos\theta_2 = \frac{\gamma_1}{2\gamma_2} = \frac{50}{2\times 70} = 0.3571$$

$$\theta_2 = 68.1^\circ$$

10.4.2 Effect of Surface Roughness on Contact Angle

Commonly, the surface of a solid is not smooth or homogeneous. Even though the surface appears smooth when observed by the naked eye, it is rough when observed with a microscope. Therefore, a roughness factor is defined as

$$r = \frac{S_r}{S_a} \tag{10.89}$$

where S_r is the real surface area and S_a the apparent surface area. If a surface is rough, the roughness factor will be larger than 1.0. For a rough surface, Young's equation Eq. (10.82) becomes

$$r(\gamma_{sA} - \gamma_{sl}) = \gamma_{lA} \cos\theta' \tag{10.90}$$

Where θ' is termed the apparent contact angle. Dividing Eq. (10.88) by Eq. (10.82) gives

$$r = \frac{\cos\theta'}{\cos\theta} \tag{10.91}$$

For the contact angle smaller than 90 degrees, since $r > 1$, we must have

$$\theta' < \theta$$

This implies that making the surface rough is favorable for wetting. Whereas, for a contact angle larger than 90 degrees, since $r > 1$, we must have

$$\theta' > 0$$

This implies that making the surface rough is not favorable for wetting.

10.4.3 Spreading of Liquid on Solid Surface

The spreading of liquid on a solid surface is considered to be the process of substituting a liquid–solid interface and a gas–liquid interface for a gas–solid interface. According to the change in free energy, the term spreading coefficient is defined as

$$S = \gamma_{sV} - (\gamma_{lV} + \gamma_{sl}) \tag{10.92}$$

where subscript V represents the vapor phase, l the liquid phase, and s the solid phase. Thus we have

$S > 0$: Free energy decrease; liquid spreads on a solid surface spontaneously
$S < 0$: Free energy increase; liquid does not spread on a solid surface spontaneously

Substituting Eq. (10.82) (Young's equation) into Eq. (10.92) gives

$$S = \gamma_{sl} + \gamma_{lV}\cos\theta - \gamma_{lV} - \gamma_{sl} = \gamma_{lV}(\cos\theta - 1) \tag{10.93}$$

To estimate the spreading coefficient of a liquid on a solid surface from Eq. (10.93), first we should know the contact angle, which is not easy to measure. A simple and easy method called droplet-height method was proposed.

Place a droplet of liquid on a solid surface and add the liquid on the top of this droplet constantly drop by drop. If the liquid does not wet the solid, a liquid column will form, and its height will increase with the increase in the amount of liquid added on the top of the droplet until the limiting height is reached. Continuing to add the liquid on the top will not increase the height of the column, and the liquid added later will drop down, thus leading to the increase in the area of the liquid–solid interface, as shown in Fig. 10.18.

Suppose the limiting height of the liquid column is h, the decrease in the height of the liquid column due to the dropping down of liquid dh, the radius of the liquid–solid interface r, and the increase in the radius of the liquid–solid interface due to the dropping down of liquid dr. Since the dropping down of liquid on the column causes increases in the areas of the gas–liquid interface

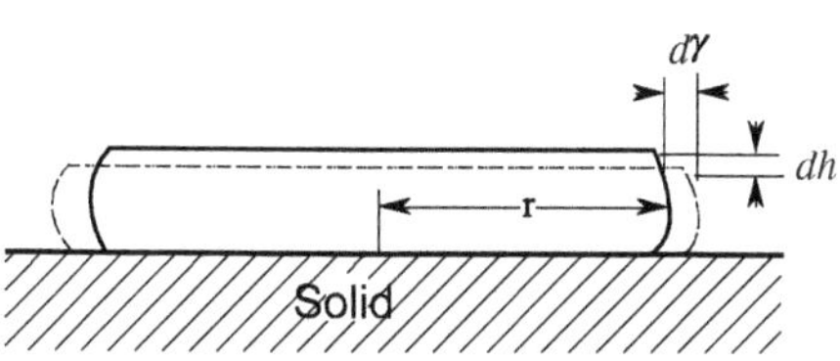

FIGURE 10.18 Measurement of spreading the coefficient by droplet-height method.

and liquid–solid interface, the corresponding increase in the surface free energy is given by

$$2\pi r dr\left(\gamma_{sl}+\gamma_{lv}-\gamma_{sv}\right) \tag{10.94}$$

Let the volume of the liquid be V, the falling height of center of gravity of the column be $dh/2$, then the decrease in potential energy is given by

$$\frac{1}{2}\rho g V\cdot dh \tag{10.95}$$

This decrease in potential energy must be transformed to the interface free energies in this process. Equating the decrease in potential energy and increase in surface free energy gives

$$2\pi r dr(\gamma_{sl}+\gamma_{lv}-\gamma_{sv})=\frac{1}{2}\rho g V\cdot dh \tag{10.96}$$

or

$$(2\pi r dr)h(\gamma_{sl}+\gamma_{lv}-\gamma_{sv})=\frac{1}{2}\rho g V\cdot dh\cdot\frac{h^2}{h} \tag{10.97}$$

Since the volumes of falling liquid before and after falling are equal, that is,

$$(2\pi r dr)h=V\frac{dh}{h} \tag{10.98}$$

Eq. (10.97) becomes

$$(\gamma_{sl}+\gamma_{lv}-\gamma_{sv})=\frac{1}{2}\rho g h^2$$

or

$$\gamma_{sv}-\gamma_{sl}-\gamma_{lv}=-\frac{1}{2}\rho g h^2$$

or

$$S=-\frac{1}{2}\rho g h^2 \tag{10.99}$$

This implies that if the limiting height is measured, the spreading coefficient will be obtained conveniently by Eq. (10.99).

Example 10.9 Spreading of Liquid on Solid Surface

The limiting height of the water column on a paraffin surface at 27.5°C is measured to be 0.435 cm, the surface tension and the density of water at 27.5°C are known as 69.2 mN m^{-1}, and 0.997 g cm^{-3}, respectively. Evaluate the spreading coefficient and contact angle of water on the surface of paraffin.

Solution

According to Eq. (10.99)

$$S = -\frac{1}{2}\rho g h^2 = -\frac{1}{2} \times 0.997 \times 981 \times (0.435)^2 = -92.5\left(\text{mN m}^{-1}\right) < 0$$

Therefore water does not spread on the surface of paraffin spontaneously.
According to Eq. (10.93)

$$S = \gamma_{1V}(\cos\theta - 1)$$
$$\cos\theta = \left(\frac{S}{\gamma_{1V}}\right) + 1 = \left(\frac{-92.5}{69.2}\right) + 1 = -0.337$$
$$\theta = 100 \text{ degrees}$$

In 1930 Osterhof and Bartell classified wetting into three types, which are adhesion, immersion, and spreading. According to this classification, the process shown in Fig. 10.13 is adhesion, the process discussed in this section is spreading, and immersion refers to the case in which a whole object is placed in liquid (under the surface of liquid), thus all the gas–solid interface disappears and is substituted by a liquid–solid interface. If the total area of the solid object is 1 unit, at a constant temperature and pressure, the change in free energy of the system is given by

$$\Delta G = \gamma_{sL} - \gamma_{sV} \tag{10.100}$$

Then the immersion work is expressed as

$$W_i = -\Delta G = \gamma_{sV} - \gamma_{sL} \tag{10.101}$$

Only if the surface tension of solid is greater than that of the liquid–solid interface, the immersion will take place spontaneously.

10.4.4 Wetting and Water Treatment

The principle of wetting is very important for water treatment. In the coarse graining method (one of the oil water treatment methods), lipophilic materials are used as the filler to make fine oil droplets condense on the hydrophobic surface. Filtration is commonly used as the method for oil water treatment in depth, in which the properties or the kinds of filter medium are the key factors, especially the wetting property of oil for filter medium, which plays a great role for oil removal. If oil pollutants wet filter medium better, the removal rate of oil pollutants of water by filtration will be higher. Therefore, studies on the contact angle and wetting property of oil for filter medium is helpful for the selection of filter medium which applies to oil water treatment. Membrane filtration including reverse osmosis, ultrafiltration, microfiltration, and nanofiltration are widely used in water treatment, for example, desalination of sea water or brackish water, preparation of pure water, and the treatment of many kinds of industrial wastewaters and domestic sewages. However, membrane pollution, which causes the blocking of membrane pores and a decrease in membrane flux, is the most serious problem in membrane filtration technology. The extent of membrane pollution is directly related to the wetting property of the membrane. The commonly used polyvinylidene fluoride membrane and polypropylene membrane are strongly hydrophobic and therefore very easy to pollute. To reduce membrane pollution, the hydrophilic property of the membrane surface should be enhanced by surface modifications.

In recent years, accidents involving leakage of crude oil on the sea are frequent, causing serious pollution of sea water. To restore the sea water, some superhydrophobic or superlipophilic materials have been developed to adsorb and collect the oil on the sea efficiently when an emergency takes place. The contact angle between water and the superhydrophobic materials is commonly larger than 150 degrees.

10.5 MEASUREMENT OF PROPERTIES OF ADSORBENTS

10.5.1 Measurement of Specific Surface Area

The total surface area per unit mass of adsorbent is termed specific surface area, and its unit commonly used is $m^2\ g^{-1}$. Specific surface area, which is directly related to the adsorption capacity, is one of the most important parameters of adsorbents.

Gas adsorption is a common method for measuring specific surface area. In this method first we measure the data of adsorption isotherms, then evaluate the monolayer adsorption capacity by using an adsorption model, then multiply it by the section area of gas molecules, thus we can obtain the specific surface area.

1. BET method of two constants
   Up to now the nitrogen adsorption method is still the standard method accepted by most researchers to measure the specific area of adsorbents. The basic procedures are: First, let the sample of adsorbent adsorb the nitrogen gas at the temperature of $-195°C$, then measure the amount of adsorption [cm^3 (STP) g^{-1}] at equilibrium for 5~8 relative pressures P/P_0 in the range of 0.05~0.35, respectively, at last making a linear fitting for the BET equation of two constants. The adsorption data of N_2 gas will give the specific area of the adsorbent. The detail of procedures is shown in Example 10.3.
2. BET one-point method
   When $c \gg 1$, the linear equation of BET, that is, Eq. (10.53), can be simplified as

$$\frac{P}{V(P_0 - P)} = \frac{1}{V_m} \cdot \frac{P}{P_0} \tag{10.102}$$

It can be seen from this equation that the plot of $P/V(P_0 - P)$ versus P/P_0 should be a straight line which passes through the original point, and the slope of this straight line is $1/V_m$. For BET one-point method, we need to measure only one amount of adsorption V at equilibrium for a pressure P, then make the point $(P/V(P_0 - P), P/P_0)$ in the coordinate system, and connect this point with the original point, thus obtaining V_m by estimating the slope of the line. In addition, Eq. (10.100) can be rearranged as

$$V_m = V\left(1 - \frac{p}{P_0}\right) \tag{10.103}$$

Therefore we can also obtain the V_m by substituting the V measured previously and the corresponding pressure P into Eq. (10.103). Then the specific area of adsorbents can be calculated as follows

$$S = V\left(1 - \frac{P}{p_0}\right)N_{\mathrm{A}}\sigma \tag{10.104}$$

where V is the amount of adsorption at equilibrium for pressure P (expressed as mol g^{-1}), N_A is the Avogadro constant, and σ the section area of one molecule of nitrogen gas.

The error of BET one-point method is generally in the range of 5% compared with the multipoint method. For the adsorbents of silica gel and aluminum oxide, generally the result obtained by BET one-point method is satisfactory. Obviously, BET one-point method is fast, simple, and easy.

10.5.2 Measurement of Pore Volume

1. Method of adsorption of carbon tetrachloride
   In this method, we use carbon tetrachloride (CCl_4) as the adsorbate. The pore volume of adsorbent sample V ($cm^3\ g^{-1}$) is evaluated by the mass of CCl_4 adsorbed into it. In the beginning of the measurement, the mixed liquid of tetrachloride and hexadecane of volume ratio of 86.9:13.1 is added into an adsorption apparatus. The relative pressure of CCl_4 (P/P_0) approximately equals 0.95 in this mixed liquid. Then the two weighing bottles with the same volumes and the same weights are placed in the adsorption apparatus. One contains the adsorbent sample, the other is empty for correcting the amount of CCl_4 adsorbed on the wall of the weighing bottle. After 16 h adsorption, which is enough to reach adsorption equilibrium, the two bottles are taken out and weighed respectively. The pore volume is calculated by

$$V = \frac{W_1 - W_2}{W_s \rho} \tag{10.105}$$

   where W_1 is the weight of the bottle containing the adsorbent sample, W_2 the weight of the empty bottle, W_s the weight of adsorbent sample, and ρ the density of CCl_4 at the temperature of measurement.

It was proven that if $P/P_0 > 0.95$, the condensation of CCl_4 between the particles of an adsorbent sample will take place, thus leading to too high of a value of pore volume.

2. Density method
If the granular density and true density of a porous adsorbent is determined, its pore volume can be calculated by

$$V = \frac{1}{d_g} - \frac{1}{d_t} \tag{10.106}$$

where d_g is the mass of the porous adsorbent of the unit volume (including the volume of the large amount of micropores in it), d_t the density of framework material of porous adsorbent, and the numerator 1 represents unit mass. This method is very simple and easy, but the adsorbents should be larger particles. If they are powders, this method will be not applicable.

10.5.3 Measurement of Mean Pore Radius

The relevant research revealed that the shapes of pores in porous adsorbents are very complicated. For example, the micrograph of the section of silica gel shows that there are large amounts of various-shaped pores including circle, ellipse, triangle, dumbbell, and irregular shapes. For simplicity, the pore is assumed to be cylindric, thus, its volume is expressed as

$$V = \pi r^2 l \tag{10.107}$$

And its inner surface area is given by

$$S = 2\pi r l \tag{10.108}$$

where r is the radius of the cross-section of the cylinder and l the length of the cylinder. Combining Eqs. (10.107) and (10.108) gives the average radius of cross-sections of cylinders as follows

$$\bar{r} = \frac{2V}{S} \tag{10.109}$$

where V is the pore volume per unit mass of adsorbent (cm^3 g^{-1}) and S is the surface area per unit mass of the adsorbent (m^2 g^{-1}). This implies that if the pore volume and the surface area of adsorbent of unit mass are determined, the average radius of cross-sections of pores will be obtained as follows

$$\bar{r} = \frac{2V}{S} \times 10^3 \text{ (nm)} \tag{10.110}$$

If the pore is assumed to be conical, the average radius will be given by

$$\bar{r} = \frac{3V}{S} \times 10^3 \text{ (nm)} \tag{10.111}$$

Sometimes, the average values of the results calculated from Eqs. (10.110) and (10.111) are used.

10.5.4 Measurement of Pore Radius Distribution

Based on the IUPAC (International Union of Pure and Applied Chemistry) standard, the pores of porous materials are classified according to the pore width (represented by d) as:

1. Micropores: $d < 2$ nm
2. Mesopores: 2 nm $< d <$50 nm
3. Macropores: 50 nm $< d$

The pore size distribution is shown by the relationship curve between pore volume and pore radius or by the relationship curve between the change rate of pore volume with pore radius and pore radius. The former is referred to as the integrating curve for pore size distribution, and the latter is referred to as the differential curve for pore size distribution. The pore size distribution is commonly measured by gas-adsorption method or mercury-pressure method. The former applies to the samples of pores of radii smaller than 10 nm, the latter applies to the samples of pores of radii larger than 10 nm.

1. Gas adsorption method

   The basis of evaluating pore radius distribution is the Kelvin equation, that is, Eq. (10.64).

   Select the datum points at the desorption branch of hysteresis loop of isotherm IV or V, and calculate the corresponding pore radii r by substituting P_r/P_0 values of the selected points into the Kelvin equation. The value of r will be the corresponding pore radius at which capillary condensation takes place for relative pressure P_r/P_0 and is termed Kelvin radius. Since there is already some liquid adsorbed on the wall of pore before the capillary condensation takes place (let the thickness of liquid adsorbed be τ), the true radius r_p should be r plus τ. Transforming the amount of adsorption at the selected points into the amount of adsorption expressed in terms of volume per unit mass, we obtain the total volume of the pores of radii smaller than r_p, which is expressed as V_r. The curve of V_r versus r is termed the integrating curve for pore size distribution. Again select the datum points separated by suitable intervals at the integrating curve, and evaluate the slopes at the different points, and then plot them versus r, which will give the differential curve for pore size distribution. Figs. 10.19 and 10.20 are the integrating curve and the differential curve of a fine silica gel respectively.
2. Mercury penetration method

   Since mercury does not wet common solids, the mercury droplet larger than the pores of the adsorbent cannot enter into the pores except for when a force is applied. Fig. 10.21 illustrates how a mercury droplet is forced to enter into a pore.

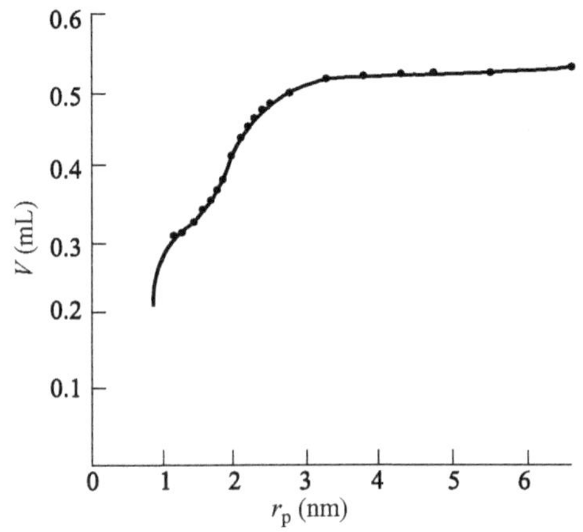

FIGURE 10.19 **Integrating curve for pore size distribution of a fine silica gel.**

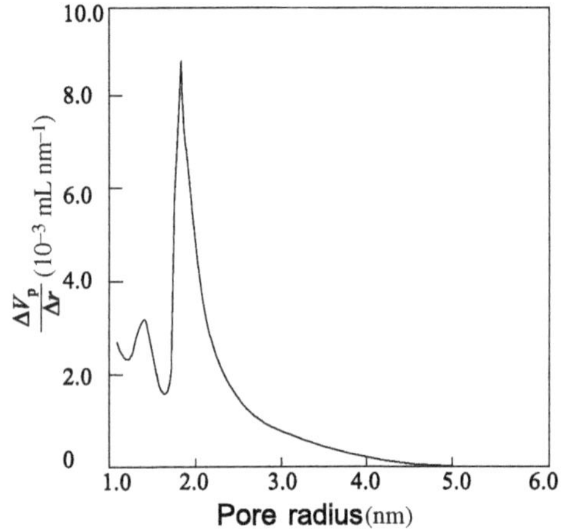

FIGURE 10.20 **Differential curve for pore size distribution of a fine silica gel.**

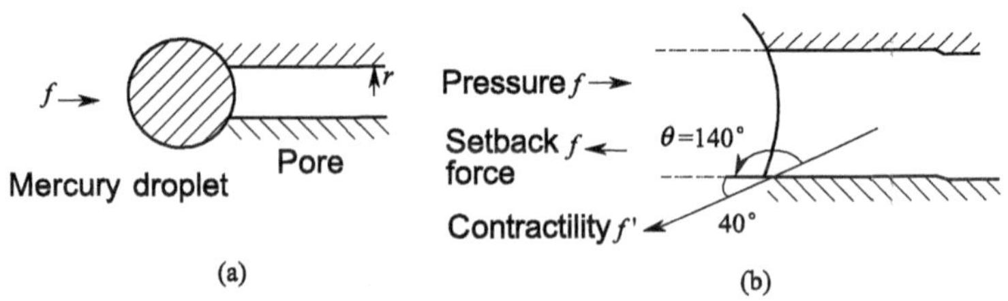

FIGURE 10.21 **Entering of a mercury droplet into the pore of an adsorbent.** (a) Mercury droplet; (b) mercury penetration.

Since the area of porthole is πR^2, the force f will be $\pi R^2 P$ where P is the pressure. If f is applied, there will be a level component of the surface tension which is of the same value as f but in the direction opposite to f. This level component of the surface tension is expressed as

$$2\pi R\gamma \cos 40° = -2\pi R\gamma \cos 140°$$

At equilibrium

$$\pi R^2 P = -2\pi R\gamma \cos 140°$$

$$\begin{aligned} R &= -\frac{2\gamma \cos 140°}{P} = -\frac{2\times 0.480\times(-0.766)}{p} \\ &= \frac{7350\times 10^5}{P} = \frac{7260}{P_{atm}}\,(\mathrm{nm}) \end{aligned} \tag{10.112}$$

where P_{atm} is the applied pressure expressed in terms of atmospheric pressure (atm) as its unit, γ is the surface tension of mercury (0.480 N m^{-1}), and θ is the contact angle of mercury on the solid (140 degrees). Eq. (10.112) is the basic equation for the mercury penetration method. It can be seen that if $P = 1.013 \times 10^5$ Pa, R will equal 7260 nm, which implies that for pores of radius 7260 nm, the pressure must be greater than 0.1 MPa to force the mercury to enter into the pores. If $P = 1013 \times 10^5$ Pa (about 1000 atm), R will equal 7.3 nm, which implies that for pores of radius 7.3 nm, the pressure must be greater than 101 MPa to force the mercury to enter into the pores. The higher the pressure, the more difficult the experiment; therefore, this method applies to measuring the porous materials of larger pores.

10.6 ANALYSIS OF THE SURFACE OF ADSORBENTS

Commonly, the surface of adsorbents is a surface layer which is one to several atoms thick. When surface analysis is carried out, a beam of photons or electrons or atoms or ions, or a probe, is directed at a surface. The surface will emit or scatter photons or electrons or ions or other neutral particles due to interaction between the surface and the beam. Determining the energy or mass-to-charge ratio or beam intensity will give a lot of surface information.

10.6.1 Analysis of the Surface Morphology of Adsorbents

Since it is not possible to observe the details of a solid surface by naked eyes or optical microscope, the electron microscope with a beam of electrons as its light source was developed.

1. Transmission electron microscope (TEM)
   TEM is a large-scale electron-optical instrument which uses the beam of electrons of high energy as its light source and an electromagnetic lens as

its magnifying glass. When the thickness of a sample is smaller than the penetrating depth of the electron beam, a part of electrons will pass through the sample and emit from the lower surface, then producing the image. If the sample is very thin, such as tens of nanometers, the main part of transmission electrons will be elastic scattering electrons, thus the image will be very clear. If the sample is thicker, the transmission electrons will be less, thus the image will be indistinct.

Since the penetrating depth of the electron beam is limited, the sample must be prepared as a thin film. For common TEM, the accelerating electric pressure of electron beam is in the range of 50~100 kV; the appropriate thickness of sample should be in the range of 1000~2000 Å, therefore, the preparation of samples is quite difficult.

2. Scanning electron microscope (SEM)
   When SEM is applied, a very fine electron beam is used to scan and bombard the surface of the sample point-by-point, thus producing the secondary electrons which will be accepted by various signal detectors and amplified as the secondary signal to be transferred to the kinescope. Since the surface of the sample is uneven, the secondary signal from the surface will change with the surface morphology. SEM reflects both surface morphology and element distribution at the same time.

SEM is a very efficient means for research of the three-dimensional structure and the surface morphology of solids. Although the resolution ratio of SEM is lower than that of TEM, it can still give a fine image for rough surfaces. In addition, it has many advantages, such as simple preparation of sample, long depth of field, third dimension, and continuously changeable magnification times, etc. The conductive sample can be measured directly by SEM, whereas the material of poor conductivity or insulator needs to spray by using conductive material to be measured by SEM.

10.6.2 Analysis of the Surface Composition of Adsorbents

The chemical composition of materials can be measured by many physical or chemical methods. The results obtained by general methods are often of average values; however, the composition of surface itself is different from the inner part of materials, that is, different from the average results. To know the composition of surface microareas, commonly we use X-ray photoelectron spectroscopy and Fourier transform infrared analysis.

1. X-ray photoelectron spectroscopy (XPS)
   When XPS is applied, a very strong beam of X-ray is used to bombard the surface of the sample, thus stimulating electrons and making them fly out of the surface and go to the energy detector. After the energy analysis, the electrons are measured, and then an X-ray photoelectron spectroscopy can be made. For a certain X-ray source, if the kinetic energies of the electrons

are determined, the binding energies of the electrons can be calculated by the corresponding equation. Since the binding energies of electrons are different for different atoms or molecules, they can be used to identify atoms and molecules. Because XPS can be used to identify the chemical states of nonmetals and the oxide states of metals, it is referred to as electron spectroscopy for chemical analysis (ESCA). The popularity of XPS stems from its following abilities:

a. To identify and quantify the composition of all elements from Li-U within 10 nm or less layer of any solid. Note: This is on the assumption that the element of interest exists at >0.05 atomic percent. (H and He are not detectable due to their extremely low photoelectron cross-section and the fact that XPS is optimized to analyze core electrons.)

b. To reveal the chemical environment where the respective element exists, that is, the speciation of respective elements observed.

c. To obtain the previously mentioned information with relative ease and minimal preparation.

2. Fourier transform infrared analysis (FTIR)

If a beam of infrared rays with various wavelengths is directed at the molecules of a sample, some rays with certain wavelengths will be absorbed, thus an infrared absorption spectrum will be obtained. Since every kind of molecule has its characteristic infrared absorption spectrum which depends on its own composition and structure, the infrared spectroscopy can be used to identify and analyze molecule structure, chemical bonds, and functional groups, and to characterize materials. The infrared absorption spectrum is produced by the unceasing vibration and turn of molecules. The energies of the vibrations of molecules correspond to the energies of quantum photons of infrared ray; therefore, the infrared ray can stimulate molecules and make them vibrate, thus producing infrared absorption spectrum. Fourier transform infrared spectrometer consists of light source, interferometer, sample cell, detector, and computer, in which the light from the light source is transformed into interference light before absorption, and the Fourier transform of interference pattern obtained after absorption produces the pattern of absorbance versus frequency. FTIR has many advantages, such as multichannel measurement, high light quantity, accurate wave number, and high resolution ratio, and therefore has been widely used for surface analysis of adsorbents in recent years.

10.7 ADSORPTION IN NATURAL WATER AND WATER TREATMENT

Adsorption of particles in natural water is a very general phenomenon. It includes physical adsorption, ion exchange, and surface complexation at the surfaces of metal oxides, hydroxides, hydrous oxides, and humic substances, etc. These surfaces are commonly made up of OH groups capable of showing weak acid behavior, and the extent to which they are protonated determines the surface charge. At a pH above the point of zero charge (pzc), the solid has a net negative surface charge, which increases with pH and electrolyte concentration. This negative charge is largely balanced by the adsorption of metal cations such as heavy metal ions. A number of researches show that most heavy metals in natural waters combine with the particles, and most persistent organic pollutants (POPs) in wastewater and polluted natural waters are very easy to combine with inorganic suspended substances and transform to deposit sediment.

Currently, adsorption is widely applied in water treatment mainly to remove the trace pollutants from water or to treat water in depth or to solve the accident caused by emergencies of water pollution. The effects of adsorption include decoloration, deodorization, removing heavy metal, and dissolving organics and radioactive substances. Activated carbon is the most common adsorbent in adsorption operation of water treatment, and many other adsorbents including activated zeolite, activated aluminum oxide, silica gel, various synthesized adsorbents, and different modified natural materials, such as chitosan, are often used in water treatment. In all adsorbents, activated carbon is the most efficient because it has a very large specific area, which reaches up to 1000 $m^2\ g^{-1}$. Activated carbon can be manufactured nearly by all materials containing carbon, for example, wood, saw dust, coal, peat, shell, and bagasse. The production process of activated carbon is divided into two procedures, carbonization and activation. Carbonization, which is also referred to as pyrolysis, is heating materials in the condition of isolation from air at a temperature below 600°C to make the materials release water vapor, carbon monoxide, carbon dioxide, hydrogen gas, etc., thus forming stable structures. Activation is heating the carbonized materials in the presence of oxidizing agents (generally, vapor is used as the oxidizing agent) at the temperature of 800~900°C. If the temperature is below 600°C, air can be used as the oxidizing agent. The effect of activation is to burn off the hydrocarbons adsorbed on the materials in the process of carbonization and the carbon atoms originally located at the edges of pores, enlarge the volume of pores, increase specific area, and give a good porous structure to the product, etc. Activated carbons are classified as particle product and powder product, which are used for different purposes. When the particle-activated carbon is used, it is commonly filled into a column as one of the units of technical processes. When powder-activated carbon is used it is often added into water directly and separated from water after adsorption. Activated carbon needs to be regenerated when it reaches its adsorption saturation after working for a certain time. The thermolysis by furnace in fluid bed and the blow wash by vapor are used for regeneration. The regeneration efficiency of the former is often above 99%, whereas the regeneration efficiency of the latter commonly is lower. The activated carbon regenerated by the latter method can be used only for several circulations of use regeneration.

The adsorption of pollutants from solution is also one of the most important mechanisms of filtration operation in water treatment. Suppose the particles of the filtration medium on the top layer of the filtration bed are spherical, and their diameter is about 0.5 mm, the pore size between the particles on the top layer will be 80 μm. Although the particle sizes of most pollutants entering the filtration bed are generally smaller than 30 μm, they can still be retained by the filtration bed in practice, even if they are in the depths of the filtration bed where the pore sizes are larger than that in the top layer of the filtration bed. This fact implies that the efficiency of filtration does not mainly depend on the mechanical retain effect. Many researchers indicated that the removal of pollutants in filtration mainly depends on the migration of pollutants between particles and the adsorption at the surface of particles, in which van der Waals force, electrostatic attraction, hydrophobic effect, coordination reaction, etc., play very important roles. Therefore, filtration may be rendered more dynamic or more efficient by modifying the chemical variables, such as adding coagulants, in such a way as to improve the efficiency of collision between particles and filter grains. Higher filtration rate may often be achieved without impairing the quality of the filtrate if a reduction in contact opportunities between suspended particles and filter grains is compensated for by a chemical improvement of the collision efficiency. Currently, many kinds of water treatment filter media of different surface properties, including nutshell, manganese sand, ceramic granule, quartz sand, anthracite, zeolite, and magnetite, are selected for different kinds of pollutants containing different properties.

Chapter 11

Emulsion, Foam, and Gel

Chapter Outline

11.1 EMULSION

11.1.1 Formation of Emulsion and Its Type

Emulsion is a dispersed system in which the phases include immiscible or partially miscible liquids, such as milk and rubber latex. If an emulsion is prepared by homogenizing two pure liquid components, phase separation will generally be rapid, especially if the concentration of the dispersed phase is at all high. To prepare reasonably stable emulsions a third component, an emulsifying agent (or emulsifier), must be present, such as surfactant or solid powder. The functions of the emulsifying agent are to facilitate emulsification and promote emulsion stability. The stabilizing mechanism is generally complex and may vary from system to system; however, two of them are very common and very important.

1. Low interfacial tension: The adsorption of surfactants at the interface of phases causes a lowering of interfacial energy, thus facilitating the development of emulsion and enhancing the stability of the large interfacial area associated with emulsion.

Colloid and Interface Chemistry for Water Quality Control. http://dx.doi.org/10.1016/B978-0-12-809315-3.00011-6

2. A mechanically strong interfacial film: The stability of emulsion arises from the mechanical protection given by the adsorbed films around the droplets, rather than from a reduction of interfacial tension.

In nearly all emulsions, one of the phases is aqueous and the other is oil (in the widest sense of the term). Thus the emulsions are grouped into two types according to the properties of phases as follows.

1. If the oil is the dispersed phase, the emulsion is termed an oil in water (O/W) emulsion.
2. If the water is the dispersed phase, the emulsion is termed a water in oil (W/O) emulsion.

There are three methods by which the emulsion type may be identified:

1. Coloring method: Add a red dye, which is soluble in oil into an emulsion, and observe it through a microscope. If the droplets are red, while the medium around them appears to be bright, the emulsion must be O/W; if the droplets appear to be bright, while the medium around them is red, the emulsion must be W/O.
2. Electrical conductivity method: Since O/W emulsions generally have much higher electrical conductivity than W/O emulsions, the emulsion type can be identified by measuring the electrical conductivity of the system.
3. Diluting method: Mix the emulsion with an oil to dilute the system. If the emulsion mixes readily with the oil, the emulsion must be W/O, otherwise it is O/W. Or mix the emulsion with water to dilute the system, if the emulsion mixes readily with the water, the emulsion must be O/W, otherwise it is W/O.

Several theories relating to emulsion type have been proposed. The most important three theories among them are introduced as follows.

1. Phase volume theory: The larger its phase volume, the more likely a liquid is to become the dispersion medium. If the emulsion consisted of an assembly of closely packed uniform spherical droplets, the volume fraction ϕ of the dispersed phase would equal 0.7402 of the total volume, and the medium 0.2598. Therefore if $\phi > 0.7402$, the type would transform. If $\phi = 0.26 \sim 0.74$, both O/W and W/O emulsions would be possible; if $\phi > 0.7402$ or $\phi < 0.2598$, only one type would be possible—O/W or W/O. Stable emulsions can, however, be prepared in which the volume fraction of the dispersed phase exceeds 0.7402, because (1) the droplets are not of uniform size and can, therefore, be packed more densely; and (2) the droplets may be deformed into polyhedrals.
2. "Oriented wedge" theory: This theory requires that (to achieve maximum interfacial density of emulsifier, thus producing the maximum decrease in free energy and the maximum increase in strength of interfacial film) the end of a surfactant emulsifier molecule, which has the greater cross-sectional

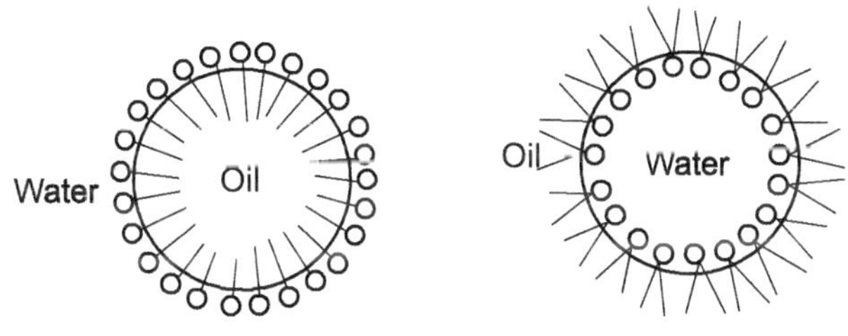

FIGURE 11.1 **Oriented wedge theory.** (a) Monovalent soaps and (b) bivalent soaps.

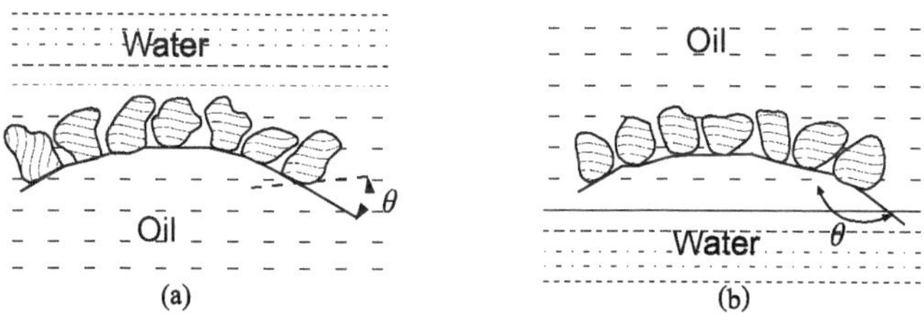

FIGURE 11.2 **Stabilization of emulsions by finely divided solids.** (a) Preferential wetting by water and (b) preferential wetting by oil.

area, should be oriented toward the dispersion medium. Thus monovalent soaps should tend to give O/W emulsion, as shown in Fig. 11.1a, and polyvalent soaps should tend to give W/O emulsion, as shown in Fig. 11.1b. In this theory the surfactant emulsifier molecules are assumed be shaped like wedges; therefore, it is termed "Oriented wedge" theory.

3. Preferential wetting theory: The most satisfactory general theory of emulsion type is that originally proposed for emulsions stabilized by finely divided solids. If the solid is preferentially wetted by one of the phases, then more particles can be accommodated at the interface if the interface is convex toward that phase, that is, if the preferential wetting phase is the dispersion medium, as show in Fig. 11.2. For example, bentonite clays (which are preferentially wetted by water) tend to give O/W emulsions, whereas carbon black (which are preferentially wetted by oil) tend to give W/O emulsions.

11.1.2 Stabilization and Breaking of Emulsion

11.1.2.1 Stabilization of Emulsion

The factors that influence the stability of emulsion include the lowering of interfacial energy and the strength of interfacial film.

Lowering of interfacial tension will result in lowering of interfacial energy, thus increasing the stability of emulsion. For example, if paroline is dispersed in water, the emulsion will be very difficult to form because the interfacial tension reaches up to 41 mN m^{-1}, but if some olive oil is added as the emulsifier, the emulsion will form spontaneously because the interfacial tension will be decreased to 0.002 mN m^{-1}.

The strength of interfacial film is also a very important factor. For example, although the alcohols of fewer carbon atoms have higher surface activity, therefore are capable of reducing the interfacial tension, the emulsions stabilized by them is not stable because the molecular chains are too short to form the thicker film to protect the droplets. However, although solid powders do not have higher surface activity, the emulsions stabilized by them is very stable because they are capable of forming stronger interfacial film to protect the droplet.

If the ionic emulsifier is used, the stability of emulsion will be strengthened because the droplets have the same charge in such cases that they are repulsive to each other.

11.1.2.2 Breaking of Emulsion

In many instances it is the breaking of an emulsion, which is of practical importance. Examples are the creaming, breaking, and inversion of milk to obtain butter, and the breaking of W/O oil-field emulsions. The breaking of emulsion includes layering, flocculation, and coalescence. The layering is caused by the difference of densities between the oil phase and the water phase, as shown in Fig. 11.3a. It is noted that the layering is only the breaking of the uniformity of emulsion; if a small quantity of electrolyte or flocculants are added to the emulsion system, the flocculation of droplets might take place and form flocs, as shown in Fig. 11.3b, but it will be recovered when the flocs are stirred; the combination or coalescence of the liquid inside the flocs will lead to thorough breaking of emulsion, as shown in Fig. 11.3c.

A number of techniques are commercially used to accelerate emulsion breakdown as follows.

1. Physical methods: Physical methods include centrifugal separation, freezing, distillation, and filtration. In addition, another method is based on the principle of antagonistic action, that is, the addition of O/W-promoting emulsifiers tends to break W/O emulsions and vice versa. Emulsions can

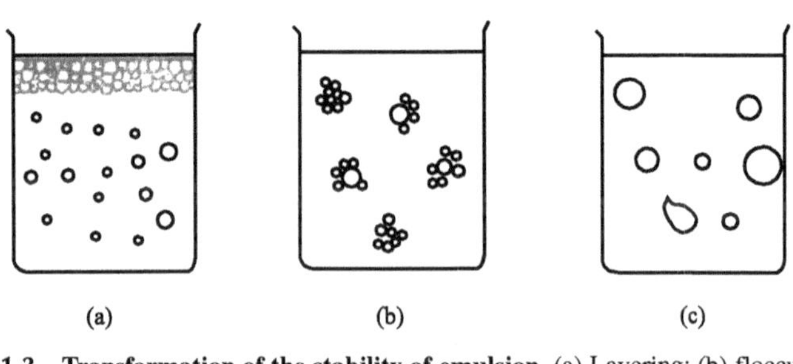

FIGURE 11.3 Transformation of the stability of emulsion. (a) Layering; (b) flocculation; and (c) coalescence.

also be broken by the application of intense electrical fields, in which the charged droplets move to the electrodes, discharge, and combine each other.

2. Physico-chemical methods: Physico-chemical methods include adding acid to the emulsion stabilized by alkali-metal soaps to change the alkali-metal soap to the acid, and adding the surfactant of short molecular chains or the surfactant with graft chains to emulsion to substitute for the surfactant of long molecular chains originally accommodated at the interface as the emulsifier.

Example 11.1 Emulsion

Give the reason why solid powder can be used as emulsifier. What types of emulsions do $CaCO_3$ powders and rosin powders form?

Solution

If solid powders can be wetted by both water and oil, they will be accommodated at the interface between water and oil, thus forming an interfacial film to protect the droplets; therefore, solid powders can be used as emulsifiers. Since $CaCO_3$ powders are preferentially wetted by water, they tend to give O/W emulsions, whereas rosin powders are preferentially wetted by oil, thus they tend to give W/O emulsions.

11.1.2.3 HLB Number

The amphiphilic nature of many emulsifying agents (particularly nonionic surfactants) can be expressed in terms of an empirical scale of the so-called HLB (hydrophile–lipophile balance) number. Generally, the numbers of HLB are in the range of 1~40. The smaller the HLB number, the more hydrophilic the emulsifying agent or surfactant; the greater the HLB number, the more lipophilic the emulsifying agent or surfactant. The emulsifying agents with HLB numbers from 3~6 are applicable to the preparation of W/O emulsions, and the emulsifying agents with HLB numbers from 8~18 are applicable to preparation of O/W emulsions. The typical applications of surfactants with different HLB numbers are listed in Table 11.1.

TABLE 11.1 Applications of HLB Numbers

HLB range	Application
3–6	W/O emulsions
7–9	Wetting agents
8–18	O/W emulsions
13–15	Detergent
15–18	Solubilizer

TABLE 11.2 HLB Numbers of Some Groups

Group	HLB	Group	HLB
$—SO_4Na$	38.7	—OH(free)	1.9
—COOK	21.1	—O—	1.3
—COONa	19.1	—OH (anhydrosorbitol)	0.5
—N (tertiary amine)	9.4	$—C_2H_4O—$	0.33
Ester (anhydrosorbitol)	6.8	$—C_3H_6O—$	0.15
Ester (free)	2.4	$=CH—, —CH_2—, —CH_3$	0.475
—COOH	2.1		

A number of different formulae have been established for calculating HLB numbers of surfactants from their composition data, and many experiment methods also have been proposed for determining HLB numbers of surfactants, for example, the method from cloud-point measurements. In addition to these methods, the HLB numbers of surfactants can be obtained by consulting the corresponding handbooks or works, or estimated out from the HLB numbers of groups, which constitute the molecule of the surfactant; the estimation equation is given by

$$\text{HLB} = 7 + \sum \text{HLB(group)} \tag{11.1}$$

The HLB numbers of some groups are included in Table 11.2.

For mixed emulsifiers, approximate algebraic additivity holds. Suppose that 70% Tween 80 (HLB 15) +30% Span80 (HLB 4.3) is the optimum composition of a mixture of these emulsifiers for preparing a particular O/W emulsion. The HLB of the mixture is, therefore, $15 \times 0.7 + 4.3 \times 0.3 = 11.8$. The theory is that an HLB of 11.8 should be optimum for the formation of this particular O/W emulsion using other emulsifier systems; for example, the optimum proportions in a mixture of sorbitan tristearate (HLB 2.1) and polyoxyethylene sorbitan monostearate (HLB 14.9) should be approximately 24.2 and 75.8%, respectively. It is clear that the optimum HLB number for the formation of a particular emulsion can be determined experimentally by using a series of mixtures with different HLB numbers, which are prepared by different pairs of surfactants of different composition ratio. Fig. 11.4 illustrates the method and result.

Example 11.2 HLB Value

The optimum HLB number for the formation of the emulsion of methylbenzene in water is 12.5. Use 4% aqueous solution of sodium oleate and 4% aqueous solution of Span20 to make the mixed emulsifier to prepare this oil in water emulsion.

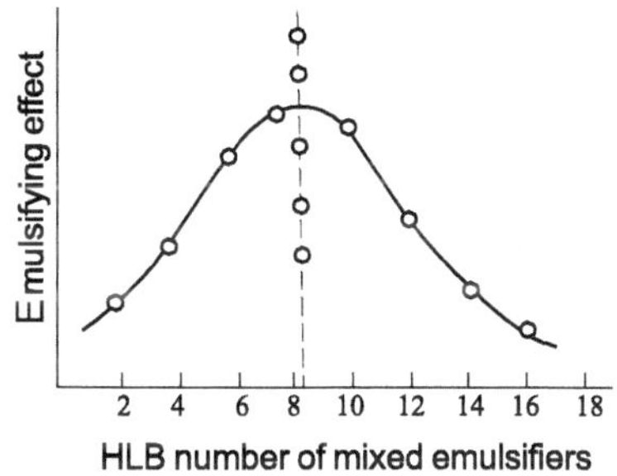

FIGURE 11.4 Determining the optimum HLB number for a particular emulsion.

How many milliliters of each of the above solutions should be taken respectively to prepare 10 cm^3 of the mixed emulsifier solution?

Solution

By consulting the corresponding handbooks we know that the HLB number of sodium oleate is 18, and the HLB number of Span20 is 8.6. Let the molar fraction of sodium oleate in the mixed emulsifier be x, thus

$$HLB_{mix} = 18x + (1-x)8.6$$

$$x = \frac{(HLB_{mix} - 8.6)}{(18 - 8.6)} = 0.42$$

V (sodium oleate) = 4.2 cm^3
V (Span20) = 5.8 cm^3

11.1.3 Applications of Emulsion in Wastewater Treatment

Oil wastewater is one of the most common industrial wastewaters. It is largely produced every day in many industries, including oil exploitation, petroleum transportation and processing, petrochemical engineering, and electromechanical engineering. The oil wastewater causes serious harm to aqueous environments due to its chroma, smell, high BOD and COD, and oxidability.

The oil in wastewater can be grouped into five types according to their different existing forms. They are floating oil (diameter d > 100 μm), dispersed oil (10 μm < d < 100 μm), emulsified oil (d < 10 μm), and dissolved oil and the oil attached to solids (d > 10 μm), in which the dispersed oil and emulsified oil are very difficult to treat due to their higher stabilities. Currently, the oil wastewater is treated by many kinds of technologies, including oil separator, air flotation, filtration, flocculation, coarse graining, membrane separation, and biochemical process. In all these technologies the principles of stability and instability of emulsion are involved and significantly influence the efficiency of

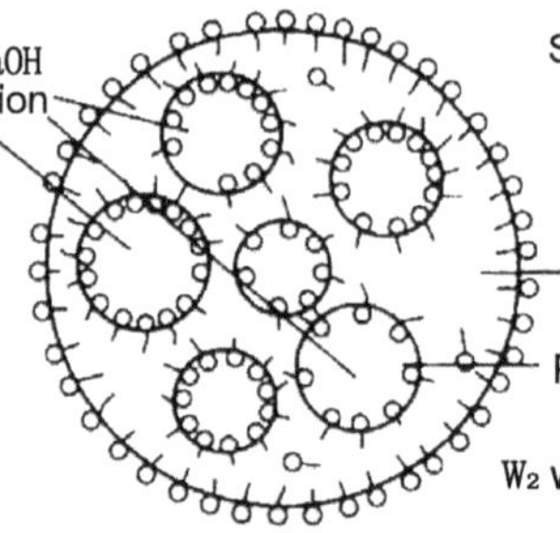

FIGURE 11.5 Schematic illustration for the removal of phenol from wastewater.

the oil removal. Therefore, studies on the mechanism of stability and instability of emulsion is important and will help us to enhance the efficiency of oil wastewater treatment and improve the technology greatly.

The liquid membrane separation technology has been developed based on the principle of emulsion formation. The typical example of liquid membrane separation in wastewater treatment is the removal of phenol of low concentration from wastewater. Firstly, let us prepare a W/O emulsion (primary emulsion), in which the water phase is 2% NaOH solution, and the oil phase contains a surfactant (98% S100N oil + 2% Span80), then disperse it in the wastewater containing phenol of low concentration, thus obtaining a W/O/W multiple emulsion, as shown in Fig. 11.5.

In Fig. 11.5 the area between the internal phase of primary emulsion and continuous external phase of multiple emulsion (wastewater) is referred to as the liquid membrane. Since phenol is partly soluble in oil, it can penetrate liquid membranes, entering the alkaline water droplet inside the liquid membrane from the water phase outside the liquid membrane, and then react with NaOH, producing sodium phenate, which is not soluble in oil and thus cannot diffuse to the phase outside the liquid membrane. As a result, the pollutant phenol is removed from the wastewater.

11.2 FOAM

11.2.1 Structure and Formation Condition of Foam

Foam is a dispersion system in which the dispersed phase is gas, and the dispersion medium is liquid or solid. If the dispersion medium is solid, it is referred to as solid foam. Foam is a coarse dispersion in which the concentration of dispersed phase is larger, and it consists of polyhedral gas cells separated by thin liquid film, that is, bubbles. Since the density difference between the dispersed phase and dispersion medium is very large, the bubbles will rise up to the surface of the liquid, forming the aggregates of bubbles separated by thin liquid films.

Only transitory foam can be formed with pure liquids and, as with emulsions, a third component—a foaming agent—is necessary to achieve any reasonable

FIGURE 11.6 Structure of foam.

degree of stability. Common foaming agents are surfactants, macromolecular polymers, proteins, and solid powders. These materials are able to decrease surface tension and strengthen interfacial film, thus increasing the stability of foam, as shown in Fig. 11.6.

11.2.2 Stability of Foam

Owing to their high interfacial area (and surface free energy), all foams are unstable in the thermodynamic sense. Two principal mechanisms are involved as follows:

1. The tendency for liquid films to drain and become thinner: It includes gravity drainage and surface tension drainage, as shown in Fig. 11.7.

   The downflow of liquid in the vertical film caused by gravity is termed gravity drainage. Since foam consists of polyhedral bubbles separated by thin liquid films, there are many Plateau borders at the intersection lines of bubbles, which form meniscuses. According to the Laplace equation, pressure difference $\Delta p < 0$ at meniscus, that is, the pressure of liquid in the Plateau borders, where the interface is curved, is less than that of liquid in the laminar film, thus the liquid in the laminar part of the film will flow into the Plateau borders. this mechanism is termed surface tension drainage.

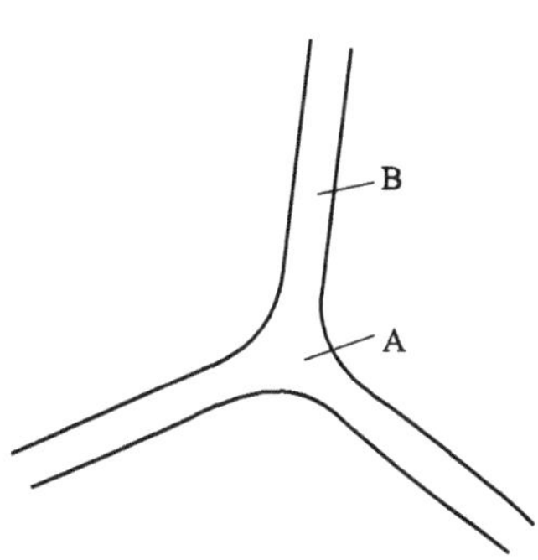

FIGURE 11.7 Gravity drainage and surface tension drainage.

2. The tendency to rupture as a result of gas diffusion: According to the Laplace equation, since the radius of curvature of smaller bubbles is smaller than that of larger bubbles, the pressure of smaller bubbles is greater than that of larger bubbles; therefore, the gas will enter into larger bubbles from smaller bubbles. Thus the small bubbles will become smaller and smaller and the large bubbles will become larger and larger. As a result, all the bubbles will disappear at last.

It was discovered that the stability of foam is influenced by the following factors:

1. It is noted that some liquids have very low surface tension, but it is not easy for them to form foam. On the contrary, although the surface activity of protein is not high, it is able to facilitate the formation of stable foams; therefore, it can be concluded that lowering the surface tension is not a very important factor for stabilizing foam.
2. The viscosity of surface film (as in the case of emulsions) often has a considerable influence on foam stability. If a little polar organic substance is added to the surfactant (foaming agent), the mixed film with high viscosity will form, which makes the drainage become difficult, thus strengthening the foam stability.
3. If a film is subjected to a local stretching as a result of some external disturbance, the consequent increase in surface area will be accompanied by a decrease in the surface excess concentration of foaming agent and, therefore, a local increase in surface tension (Gibbs effect), leading to $\gamma_A > \gamma_B$ and a surface tension gradient, as shown in Fig. 11.8. The surface pressure ($\pi = \gamma_A - \gamma_B$) will make the solution flow to spot A from spot B to cause the disturbed film region to recover its original thickness and restore the original surface tension, just like it has elasticity. This effect of automatic restoration is referred to as the Marangoni effect.
4. If an ionic surfactant is used as a foaming agent, the liquid film will be charged; therefore, the overlapping of similarly charged electric double layers opposes film thinning, thus strengthening the stability of foam, whereas nonionic surfactants can not stabilize foam very well due to lack of electrical factors.
5. If the molecules of surfactants have branched chains, commonly, their surface activity will be higher, and their diffusion coefficient will be

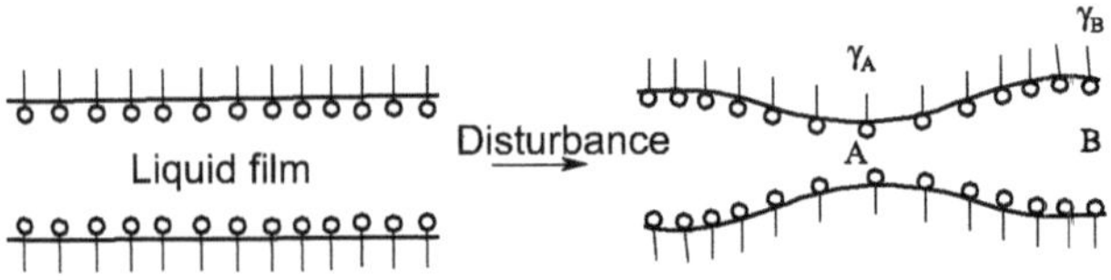

FIGURE 11.8 **"Marangoni" effect.**

greater, thus leading to strong foaming ability. However, the structure of branched chains is not favorable for the formation of strong film. If surfactants of molecules with branched chains are used, the mechanical strength of film will be smaller and the stability of foam will be lower. If the molecules of surfactants are linear, the strength of film will increase with the increase in the length of the molecular chain. But if the molecular chain is too long, the film will be too rigid; as a result, the film will lose its elasticity. In addition, the solubility of the surfactant will be too low in such cases. Generally speaking, the optimum number of carbon atoms in the molecular chains of surfactants is in the range of 12–14 for stabilizing foam.

11.2.3 Destruction of Foam

Foams are produced in the production processes of many industries, for example, soap manufacture, sugar processing, and the treatment process of starch wastewater. Foam causes a lot of problems for industrial production; therefore, the prevention of foaming or the destruction of existing foams is often a matter of practical importance. Foam inhibitors and antifoaming agents act against the various factors that promote foam stability (described earlier), thus eliminating foams.

Foam inhibitors are, in general, materials that tend to be adsorbed in preference to the foaming agent, yet do not have the requirements to form stable foam. They may be effective by virtue of rapid adsorption; for example, the addition of tributyl phosphate to aqueous sodium oleate solutions significantly reduce the time required to reach equilibrium surface tension, thus lessening the Marangoni surface elasticity effect and the foam stability. They may also act, for example, by reducing electric double layer repulsion or by facilitating drainage by reducing hydrogen bonding between the surface film and the underlying solution.

Foam can often be broken by spraying with small quantities of substances such as ether and *n*-octanol. As a result of their high surface activity, these foam breakers raise the surface pressure over small regions of the liquid films and spread from these regions displacing the foaming agent and carrying with them some of the underlying liquid. Small regions of film are, therefore, thinned and left without the properties to resist rupture.

In summary, antifoaming agents can reduce the surface viscosity of film, enhance the velocity of drainage, and raise the velocity of diffusion of gas, thus making the stability characteristic disappear.

In addition to antifoaming agents, changing pH value, adding chemicals to react with the foaming agent, and salting out can destroy foam. Some physical methods are also applied to remove foam such as temperature change, evaporation, freezing, sharp pressure change, centrifugal separation, ultrasonic wave, and filtration. Although there are lots of methods to remove foams, the destruction of foam in wastewater treatment is still quite a difficult task and should be investigated further.

Table 11.3 shows the types and applications of antifoaming agents.

Example 11.3 Emulsion and Foam

What is the difference between emulsion, foam, suspended substances, and colloids? Explain the reason we study emulsion, foam, and suspended substances as the contents of colloidal chemistry.

Solution

Generally speaking, emulsion, foam, and suspended substances are systems of low dispersion degree, their particle sizes are much greater than that of colloidal systems, which are in the range of 1~1000 nm. However, all of them are heterogeneous, unstable with respect to aggregation, and of significant interface phenomena; therefore, we study emulsion, foam, and suspended substances as the contents of colloidal chemistry.

TABLE 11.3 Types and Applications of Antifoaming Agents

Type	Name	Application
Mineral oil	Liquid paraffin	Papermaking
Oil	Animal oil, sunflower seed oil, colza oil, sesame oil	Food processing, fermentation
Aliphatic ester	Glycol distearate, sorbitol laurate, divinyl diethanol laurate, natural wax, glycerol ricinate	Paper pulp, adhesion agent, painting, fermentation, petroleum refining, dyeing, boiler water, food processing
Alcohol	Octanol, hexyl alcohol, glycol, diisobutyl carbinol, cyclohexanol	Papermaking, dyeing, painting, fermentation
Amide	Ethanediamine distearyl, didecyl amine distearyl, ethylene diamine dipalmitoyl	Boiler water, papermaking
Phosphate	Tributyl phosphate, trioctyl phosphate	Casein, fiber
Metallic soap	Aluminum stearate, calcium dioleate, potassium oleate	Lubricating oil, fiber, papermaking
Organosilicon	Dimethicone, silicon grease, modified polysiloxane, fluorosilicon oil	Food processing, fermentation, lubricating oil, papermaking, painting, adhesion agent, petroleum industry, chemical industry

Example 11.4 Foam

Describe the mechanism of eliminating foam by lowering the surface tension and forming hydrogen bonds.

Solution

1. Antifoaming agents have high surface activity that can lower the surface tension of local film when its micro droplet contacts liquid film. This local area of low surface tension will extend all around, losing elasticity and self-restoring ability, thus causing the rupture of liquid film.
2. If the surfactant that forms hydrogen bonds in the film is replaced by a surfactant that cannot form hydrogen bonds, the surface viscosity will be lowered, leading to increases in the velocity of drainage and the velocity of diffusion of gas, thus causing the rupture of liquid film.

11.2.4 Application of Foam in Wastewater Treatment

As described in Section 3.3, suspended materials can be removed from water by the sedimentation method, but the settling velocity depends upon the diameter and density of particles. If the particles are very fine and their densities are approaching or smaller than the density of water, the settling velocity will be too small to have the effect in practice; for example, the algae, plant residue in natural water, dye particle, short fiber, oil droplet, and activated sludge in wastewater are difficult to treat by the sedimentation method. To remove these pollutants from water the air flotation is developed. When it is conducted, an air compressor or vacuum pump is used to generate micro air bubbles (forming foam). These micro bubbles can adsorb the fine particles or droplets and carry them up to the surface of water, thus leading to the separation of pollutants from water. For example, the rising velocity of micro oil droplets is only about 1 $\mu m\ s^{-1}$, whereas the rising velocity of air bubbles can reach 1 mm s^{-1}, which is 1000 times the rising velocity of oil droplets; therefore, air flotation can be used to treat oil wastewater. Currently, air flotation has been successfully used in the treatment of the water supply to remove algae, plankton, low turbidity, etc. It is also used in wastewater treatment to separate or recover the surfactant, oil, heavy metal ions, pulp, activated sludge, etc.

The air flotation operation consists of two stages including gas–solid adsorption and solid–liquid separation. As far as the gas–solid adsorption is concerned, the surface tension, surface free energy, contact angle, and wetting theories must be considered. The more hydrophobic the suspended particle, the more easily they adhere to the micro air bubbles. If the contact angle between the suspended particle and water medium is smaller than 90 degrees, it will not be easy for the adhesion to take place. In this case, we commonly add a kind of surfactant to the system, changing the surface property of the particles to become hydrophobic. Since the flocs are partly hydrophobic, we often add flocculant to produce micro flocs in air flotation treatment. For the soluble pollutants such as heavy metals, first we should transform them to insoluble solids by adding some chemicals, then collect them at the surface of water by air flotation.

In some cases, the destruction of foam is needed; for example, there are large amounts of foam in starch wastewater produced in potato processing, which causes a great difficulty for the treatment of starch wastewater. It is necessary to decrease the stability of foam and break it before treatment.

11.3 GEL

11.3.1 Basic Concepts

The three-dimensional net structure formed by connected colloidal particles or molecules of polymers is referred to as gel, which is a colloidal dispersion. There are lots of gels in nature, life, and production. For example, the solution of gelatin becomes a jellylike substance when frozen; the mud for well drilling loses its original mobility and becomes a thick paste after standing for some time, bean curd is produced when brine is added to soybean milk, and silica gel is produced when acid is added to sodium silicate. Although the previously mentioned four systems are different, the changes of external conditions result in the transformations of systems from colloidal dispersions or solutions to special semisolids, and their net structures are full of liquid (dispersion medium). Two results stem from the definition of gel:

1. Gel is very different from colloidal dispersion and true solution: The particles in dispersion systems and molecules in solution are independent units which move freely; therefore, colloidal dispersion and true solution have mobility. But inside gel the particles of the dispersed phase connect one another, forming the structure in the whole system, making the system not only lose its mobility but also show the solid mechanic properties.
2. Gel is different from true solid: it consists of both solid phase and liquid phase, belonging to colloidal dispersion, therefore has limited structure strength and is easy to break so that the irreversible shape changes easily take place, thus leading the flow.

Gels are classified as nonelastic gel and elastic gel.

1. Nonelastic gel: Most inorganic gels are nonelastic gels, such as SiO_2, TiO_2, Fe_2O_3, etc. Nonelastic gels have lower mobility and smaller volume changes when they absorb or release liquid. They have no selectivity for the liquids to be absorbed and do not expand.
2. Elastic gel: The gels formed by linear macromolecules are elastic gels, such as rubber, gelatin, agar, etc. Elastic gels have higher mobility and larger volume changes when they absorb or release liquid. They have selectivity for the liquids to be absorbed and expand easily. Their shape can be restored after deformation.

Also gels can be grouped into jelly and xerogel according to the amount of liquid contained in them. Generally, jelly contains more water, which frequently reaches up to above 90%; for example, nearly 99.8 %of weight in agar is water. Jelly is generally made of flexible macromolecules and has elasticity. Xerogel contains little liquid (generally less than that in common solids). Gelatin, gum, and the semipermeable membranes used in water treatment are xerogels. The macromolecule xerogel will become jelly when it absorbs some appropriate liquid.

11.3.2 Structure of Gel

Four situations for three-dimensional net structures of gels are show in Fig. 11.9.

1. Spherical particles connect one another, for example, TiO_2 gel (Fig. 11.9a).
2. Platelike or rodlike particles connect one another, for example, V_2O_5 gel, clay slurry, and graphite slurry (Fig. 11.9b).
3. Linear macromolecules attract one another, and some of them orderly arrange, forming microcrystalline domain, for example, gelatin (Fig. 11.9c).
4. Linear macromolecules connect one another by chemical bonds, forming net structure (Fig. 11.9d).

The shapes of particles have very important effect on the minimum particle-concentration required for the formation of gel. The more asymmetrical the particle, the lower the particle concentration required for the formation of gel. The minimum volume fraction required for spherical particles to form three-dimensional net structure is 0.056. Whereas the minimum volume fraction required for many asymmetric particles to form three-dimensional net structure is much lower than this value, for example, it is 0.002 and 0.00005 for agar and V_2O_5, respectively.

The properties of particles also have very important effect on the properties and structures of gel. Flexible macromolecules form elastic gel, and rigid particles form rigid gel.

The nature of acting forces between particles has very significant effects on gel. The situations are grouped into three types as follows.

1. The structure formed by van der Waals force: This kind of structure is not stable and easy to break when external force is applied because the attraction force is weak, but can be restored by standing for some time, that is, it has thixotropic property. For example the gels of clay, graphite, $Fe(OH)_3$, etc.,

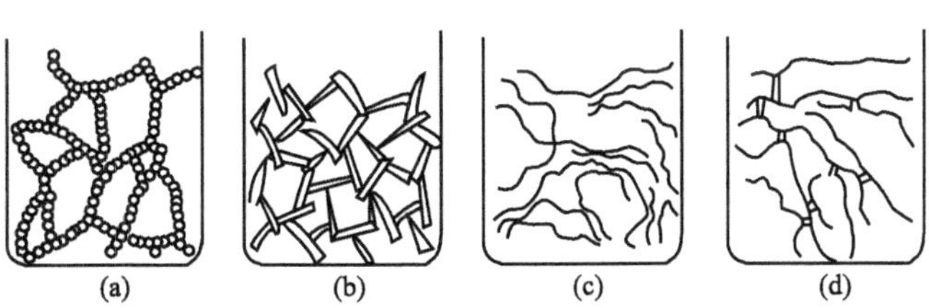

FIGURE 11.9 Structure of gel. (a) Spherical particles connect one another; (b) platelike or rodlike particles connect one another; (c) linear macromolecules attract one another; (d) linear macromolecules connect one another by chemical bonds.

are of this kind of structure. Some gels formed by linear macromolecules also are of this kind of structure, for example, unvulcanized rubber and polystyrene are of this kind of structure. When they absorb liquid, the structure will be broken down, expand, and form a solution at last.

2. The structure formed by the hydrogen bonds between macromolecules: This kind of structure is more stable than the first one. Generally, protein is of this kind of structure, such as gelatin. At low temperatures they can make limited expansions; at high temperatures they can make infinite expansions.
3. The structure formed by the chemical bonds between molecules: This kind of structure is very stable. When heated, it only exhibits limited expansion, such as silica gel and vulcanized rubber.

11.3.3 Expansion of Gel

The expansion of gel refers to the phenomenon of the increase in the volume of gel due to the absorption of liquid. It is a proprietary property of elastic gels. The expansions of gel are classified as infinite expansion and limited expansion.

1. Infinite expansion

   In this case, gel absorbs liquid at first, then its volume constantly increases and becomes a solution at last; for example, raw rubber expanding in benzene has infinite expansion. The infinite expansion of macromolecules is considered to be the first stage of dissolution and referred to as swelling.
2. Limited expansion

   In this case, gel absorbs liquid to increase its volume but not transform to sol; therefore, this kind of expansion is referred to as limited expansion. For example, the expansion of gelatin in cold water and the expansion of vulcanized rubber in benzene are limited expansions. The extent of expansion depends on the connecting strength of the structure. The limited expansion can be transformed to the infinite expansion by increasing the temperature or replacing the component of solvents. For example, gelatin exhibits limited expansion at room temperature, but when heated to above 40°C, it will dissolve completely; alternatively, if the solvent is replaced by 2 mol dm^{-3} aqueous solution of KI, it will exhibit infinite expansion at room temperature and dissolve completely at last.

The expansion of gel goes through two stages. In the first stage the molecules of solvent enter into gel and interact with macromolecules, forming the salvation layer. Generally, the change in the first stage takes place very quickly and is characterized as follows:

1. The vapor pressure of the liquid is very low, the volume contracts, that is, the increase in the volume of gel is less than the volume of the liquid absorbed. This implies that the liquid combines with macromolecules strongly.
2. The expansion is accompanied by heat release; with the increase in expansion extent the differential swelling heat (the heat released for absorption of

1 g of liquid) decreases quickly. The maximum value of differential swelling heat of various gels approximately equals the dissolution heat of molecules of low molecular weight.

In the second stage the liquid permeates into gel, increasing the volume of gel greatly. The amount of liquid absorbed is several times the amount of macromolecules themselves and even more. There is nearly no heat release, heat absorption, and volume contraction in this stage. The liquid absorbed in this stage is very easy to release when dried.

11.3.4 Diffusion in Gel

Gel can also be used as a diffusion medium. The diffusion rate of small molecules in the gel of low concentration is nearly the same as that in liquid. With the increase in the concentration of particles in gel, the diffusion rate of molecules decreases. The diffusion coefficient becomes much lower in the gel of higher crosslinking degree. On the other hand, with the increase in the size of molecules, the diffusion rate decreases that the diffusion coefficient of macromolecules is lowered significantly. The sieving action of the framework of gel makes the diffusion-rate differences due to the differences of molecule size become more significant. This sieving action can be used to separate macromolecules. In the 1960s, two important experimental technologies, gel chromatography and gel electrophoresis, were developed based on this principle. In gel chromatography, the chromatographic column is filled with gels. When the experiment is carried out, the small molecules in the sample enter the pores easily, but it is difficult for the large molecules in the sample to enter the pores; therefore, the larger molecules flow out from the column prior to the smaller molecules when eluted by solvent. By the relationships between the volume of elution liquid and concentration of molecules in the elution liquid and compared with standard samples, we can obtain the distribution curve for molecular weight easily.

Example 11.5 Gel Expansion

The change in the amount of water absorbed by gelatin with pH value is illustrated in the next figure. Give the reason why the gelatin absorbs water, that is, expands in the way the figure shows.

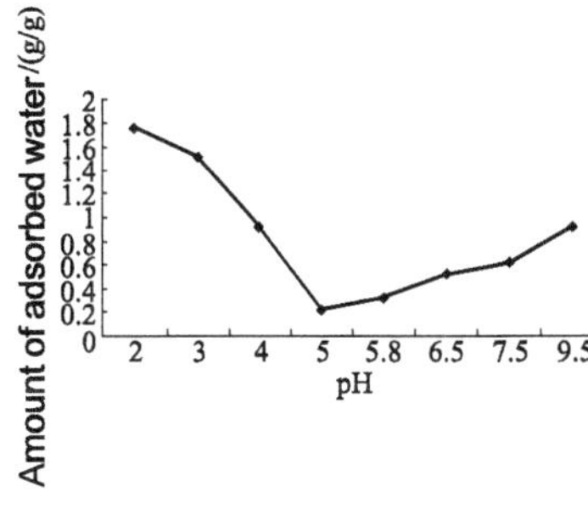

Solution

It can be seen that the minimum amount of adsorbed water is at pH 5.0, which is just the isoelectric point of gelatin. The molecules of gelatin do not ionize at its isoelectric point. Accordingly, the small ions inside the net structure of gelatin are few, therefore, no osmotic pressure is produced, thus the solvent does not penetrate toward the gelatin, that is, no gel expansion takes place. If the pH value is changed to be greater or lower than the isoelectric point, the carboxyl groups and the amino groups of gelatin molecules will ionize, producing many small ions, thus causing the osmotic pressure. Therefore, the solvent will penetrate toward the gelatin, leading to gel expansion.

11.3.5 Gels in Water Treatment

With the development of water and wastewater treatment, a large amount of sludge is producing every day in the plant of water supply and sewage treatment works. Waste biological sludge from secondary treatment is often less than 2% suspended solids. Such low solid content coupled with an increasing difficulty in obtaining permits for land disposal is forcing some industry plants to dewater their biological sludge prior to disposal. Therefore, dewatering becomes very important and necessary tasks both in water treatment plants and sewage treatment works. Currently, the optimum dewatering methods are pressure filtration, vacuum filtration, and centrifugation. In essence, the flocs sludge produced in water plants, the activated sludge produced in sewage treatment works, and the digested sludge are all gels. Therefore the properties of gels significantly influence the work of sludge dewatering. Prior to the actual dewatering process various methods of conditioning can be employed to thicken the sludge and improve its dewatering characteristics. The most commonly used method is to add chemical coagulant or flocculant to enlarge the pores in the sludge, thus promoting the release of water. The common coagulants used for sludge conditioning are aluminum sulfate, iron salts, and polyelectrolytes such as PAM.

Since the 20th century, the technologies of membrane separation including electrodialysis, reverse osmosis, ultrafiltration, microfiltration, nanofiltration, etc., have been applied to a wide variety of water treatment. In all these technologies, the key unit is the semipermeable membrane, which is gels or xerogels in essence, such as acetate fiber membrane, polypropylene membrane, and polyvinylidene fluoride membrane. The sieving action of the framework of semipermeable membrane plays the main role in membrane separation. The molecules of size smaller than pore size are able to pass through the pores of semipermeable membranes freely, but the molecules of size greater than pore size are retained. The pore size can be adjusted and controlled according to our need in the preparation of semipermeable membrane. If the semipermeable membrane is charged, it will have selectivity for permeable ions. The anions are able to pass through the membrane of positive charges, whereas the cations are able to pass through the membrane of negative charges.

In summary, the study of gels is very necessary for engineers and for students who are majoring in water treatment.

Appendix A

TABLE A.1 Basic SI Units

Physical quantity	Unit name	Unit symbol
Length	Meter	m
Mass	Kilogram	kg
Time	Second	s
Electric current	Ampere	A
Thermodynamic temperature	Kelvin	K
Luminous intensity	Candela	cd
Amount of substance	Mole	mole
Plane angle	Radian	rad
Solid angle	Steradian	sr

TABLE A.2 Derived SI Units

Physical quantity	Unit name	Unit symbol	Expression
Frequency	Hertz	Hz	s^{-1}
Force	Newton	N	$kg\ m\ s^{-2}$
Stress	Pascal	Pa	$N\ m^{-2}$
Energy	Joule	J	N m
Power	Watt	W	$J\ s^{-1}$
Charge	Coulomb	C	A s
Electrical potential	Volt	V	$W\ A^{-1}$
Capacitance	Farad	F	$C\ V^{-1}$
Resistance	Ohm	Ω	$V\ A^{-1}$
Conductance	Siemens	S	$A\ V^{-1}$
Magnetic flux	Weber	Wb	V s
Magnetic induction	Tesla	T	$Wb\ m^{-2}$
Electrical inductance	Henry	H	$Wb\ A^{-1}$
Luminous flux	Lumen	Im	cd cr
Illuminance	Lux	Lx	$Im\ m^{-2}$
Radioactivity	Becquerel	Bq	s^{-1}

Colloid and Interface Chemistry for Water Quality Control. http://dx.doi.org/10.1016/B978-0-12-809315-3.00012-8

TABLE A.3 Relation of SI Units to Other Common Units

1 m = 10^9 nm = 10^{10} Å
1 N = 10^5 dyn
1 Pa = 10 dyn cm^{-2} = 7.501 × 10^{-3} mmHg (Torr) = 9.869 × 10^{-6} atm
1 J = 10^7 erg = 0.2390 cal
1 Pa s = 10 P = 10^3 cP

TABLE A.4 Unit Prefixes

Multiples and submultiples	Prefix	Symbol
10^{12}	tera	T
10^9	giga	G
10^6	mega	M
10^3	kilo	K
10^2	hecto	H
10	deca	Da
10^{-1}	deci	D
10^{-2}	centi	C
10^{-3}	milli	M
10^{-6}	micro	μ
10^{-9}	nano	n
10^{-12}	pico	p
10^{-15}	femto	f
10^{-18}	atto	a

Appendix B

TABLE B.1 Physical Constants

Name	Symbol	Value	SI unit	CGS unit
Speed of light	c	2.99792458	10^{8} m s^{-1}	10^{10} cm s^{-1}
Vacuum permittivity	ε_0	8.85418782	10^{-12} F m^{-1}	
Electron charge	e	1.6021892	10^{-19} C	
Atomic mass unit	u	1.6605655	10^{-27} kg	10^{-24} g
Electron rest mass	m_e	0.9109534	10^{-30} kg	10^{-27} g
Proton rest mass	m_p	1.6726485	10^{-27} kg	10^{-24} g
Bohr magneton	μ_B	9.274078	10^{-24} J T^{-1}	10^{-21} erg G^{-1}
Avogadro number	N_A	6.022045	10^{23} mol^{-1}	10^{23} mol^{-1}
Boltzmann constant	k	1.380662	10^{-23} J K^{-1}	10^{-16} erg °C^{-1}
Faraday constant	F	9.648456	10^{4} C mol^{-1}	
Planck constant	h	6.626176	10^{-34} J s	10^{-27} erg s
Gas constant	R	8.31441	J K^{-1} mol^{-1}	erg °C^{-1} mol^{-1}
Gravitational acceleration	g	9.80665	m s^{-2}	10^{2} cm^{2} s^{-2}

TABLE B.2 Surface Tension of Common Liquids

Liquid	Temperature (°C)	Surface tension (dyn cm^{-1})
Isopentane	20	13.7
Diethyl ether	20	17.1
n-Hexane	20	18.4
Ethanol	20	22.3
Carbon tetrachloride	20	26.7
Chloroform	20	27.1
Acetic acid	20	27.6

(*Continued*)

Colloid and Interface Chemistry for Water Quality Control. http://dx.doi.org/10.1016/B978-0-12-809315-3.00013-X

TABLE B.2 Surface Tension of Common Liquids (*cont.*)

Liquid	Temperature (°C)	Surface tension (dyn cm^{-1})
Methylbenzene	20	28.4
Benzene	20	28.9
Carbon bisulfide	20	32.3
Methane acid	20	37.6
Phenol	20	40.9
Aminobenzene	20	44.0
Glycerine	20	63.0
Water	20	72.75
Sodium chloride	1000	98.0
Silver chloride	452	125.5
Mercury	0	470
Silver	970	800
Copper	1130	1100

Table B.3 Physicochemical Constants of Water

Term	Value	Term	Value
Molecular weight	18.015	**Conductivity**	
Freezing point	0.00°C	0°C 1.5×10^{-2} µS cm^{-1}	1.5×10^{-8} Ω^{-1} cm^{-1}
Boiling point	100.00°C		
Temperature of max density	3.981°C	18°C 4.3×10^{-2} µS cm^{-1}	4.3×10^{-8} Ω^{-1} cm^{-1}
Density (g cm^{-3})		25°C 6.2×10^{-2} µS cm^{-1}	6.2×10^{-8} Ω^{-1} cm^{-1}
Ice			
−20°C	0.9403	50°C 18.7×10^{-2} µS cm^{-1}	18.7×10^{-8} Ω^{-1} cm^{-1}
−10°C	0.9186		
0°C	0.9167	**Permittivity**	
Water		0°C	88.2
0°C	0.9999		
3.98°C	1.0000	10°C	84.3
100°C	0.9584	20°C	80.4
Vapour		25°C	78.5
100°C (1 atm)	0.5974×10^{-3}	30°C	76.8

TABLE B.3 Physicochemical Constants of Water (*cont.*)

Term	Value
Thermal conductivity	
0°C 0.502 W (m K)$^{-1}$	0.00120 kcal (cm s °C)$^{-1}$
20°C 0.599 W (m K)$^{-1}$	00143 kcal (cm s °C)$^{-1}$
75°C 0.645 W (m K)$^{-1}$	0.00154 kcal (cm s °C)$^{-1}$
Viscosity (Pa s^{-1}×10^{-4})	
0°C	17.94
20.2°C	10.00
25°C	8.94
Heat of fusion	
79.67 cal g^{-1}	333.56 J g^{-1}
Evaporation heat (100°C)	
539.0 cal g^{-1}	2256.7 J g^{-1}
Refractive index	
20°C	1.33330
Specific heat capacity	
0°C 4.2174 J (g K)$^{-1}$	1.0073 cal (g °C)$^{-1}$
14°C 4.1868 J (g K)$^{-1}$	1.0000 cal (g °C)$^{-1}$
100°C 4.2157 J (g K)$^{-1}$	1.0069 cal (g °C)$^{-1}$

Term	Value
Surface tension	
0°C	75.83 mN m^{-1} (dyn cm^{-1})
20°C	72.75 mN m^{-1} (dyn cm^{-1})
100°C	58.80 mN m^{-1} (dyn cm^{-1})
Saturated vapor pressure (1 atm)	
−10°C	0 × 101.3 kPa
0°C	0.0060 × 101.3 kPa
100°C	1.0000 × 101.3 kPa
Critical constants	
Temperature	374.2°C
Pressure	22.12 MPa
Density	0.324 g cm^{-3}
Triple point	
Temperature	0.0098°C
Pressure	610.477 Pa

TABLE B.4 HLB Value of Some Surfactants

Trade name	Type	HLB	Trade name	Type	HLB
Span 85	N	1.8	Arlacel 85	N	1.8
Span 65	N	2.1	Emcol EO-50	N	2.7
Emcol PO-50	N	3.4	Arlacel 83	N	3.7
Span 80	N	4.3	Span 60	N	4.7
Emcol DP-50	N	5.1	Atlas G-2124	N	6.1
Span 40	N	6.7	Atlas G-2147	N	7.7
Atlas G-3608	N	8	Span 20	N	8.6
Emulphor VN-430	N	9	Brij 30	N	9.5
Tween 61	N	9.6	Tween 81	N	10.0
Tween 65	N	10.5	Tween 85	N	11
Atlas G-3300	A	11.7	Igepal CA-630	N	12.8
Atlas G-1794	N	13.3	Tween 21	N	13.3
AOT	A	14	Tween 60	N	14.9
Tween 80	N	15	Tween 40	N	15.6
Tween 20	N	16.7	Atlas G-263	C	25~30
Pluronic L31	N	3.5	Pluronic L61	N	3
Pluronic L81	N	2	Pluronic L42	N	8
Pluronic L62	N	7	Pluronic L72	N	6.5
Pluronic L63	N	11	Pluronic L64	N	15
Pluronic L68	N	29	Pluronic L88	N	24
Pluronic L108	N	27	Pluronic L35	N	18.5

Appendix C

TABLE C.1 Common Equations

1. Differential

$y=a$	$\frac{dy}{dx}=0$
$y=au$	$\frac{dy}{dx}=a\frac{du}{dx}$
$y=x^n$	$\frac{dy}{dx}=nx^{n-1}$
$y=u^n$	$\frac{dy}{dx}=nu^{n-1}\frac{du}{dx}$
$y=uv$	$\frac{dy}{dx}=u\frac{dv}{dx}+v\frac{du}{dx}$
$y=\frac{u}{v}$	$\frac{dy}{dx}=\frac{v\frac{du}{dx}-u\frac{dv}{dx}}{v^2}$
$y=e^x$	$\frac{dy}{dx}=e^x$
$y=a^x$	$\frac{dy}{dx}=a^x\ln a$
$y=\ln x$	$\frac{dy}{dx}=\frac{1}{x}$
$y=e^u$	$\frac{dy}{dx}=e^u\frac{du}{dx}$
$u=f(x,y)$	$du=\left(\frac{\partial u}{\partial x}\right)_y dx+\left(\frac{\partial u}{\partial y}\right)_x dy$

2. Integral

$$\int dx = x + c$$

$$\int x^n dx = \frac{x^{n+1}}{n+1} + c$$

(*Continued*)

Colloid and Interface Chemistry for Water Quality Control. http://dx.doi.org/10.1016/B978-0-12-809315-3.00014-1

TABLE C.1 Common Equations (*cont.*)

$$\int \frac{1}{x} dx = \ln x + c$$

$$\int \frac{dx}{x+a} = \ln(x+a) + c$$

$$\int e^x dx = e^x + c$$

$$\int a^x dx = \frac{a^x}{\ln a} + c$$

$$\int \ln x \, dx = x \ln x - x + c$$

$$\int au \, dx = a \int u \, dx$$

$$\int (u+v) dx = \int u \, dx + \int v \, dx$$

$$\int u \, dv = uv - \int v \, du$$

3. Series

$$\ln(1+x) = x - \frac{1}{2}x^2 + \frac{1}{3}x^3 - \frac{1}{4}x^4 + \cdots$$

$$\ln(1-x) = -\left(x + \frac{1}{2}x^2 + \frac{1}{3}x^3 + \frac{1}{4}x^4 + \cdots\right)$$

$$e^x = 1 + x + \frac{x^2}{2!} + \frac{x^3}{3!} + \cdots$$

$$\ln x = \frac{\lg x}{\lg e} = 2.303 \lg x$$

$$e = 2.71827$$

$$(1+x)^n = 1 + nx + \frac{n(n-1)}{2!}x^2 + \frac{n(n-1)(n-2)}{3!}x^3 + \cdots$$

$$(1-x)^n = 1 - nx + \frac{n(n-1)}{2!}x^2 - \frac{n(n-1)(n-2)}{3!}x^3 + \cdots$$

$$\sum_{i=1}^{\infty} x^i = \frac{x}{1-x}$$

Further Reading

Becher, P., 1978. Emulsions, Theory & Practice, second ed. China Scientific Press, Beijing, (in Chinese).

Chang, Q., 2011. Flocculation in Water Treatment, second ed. China Chemical Industry Press, Beijing, (in Chinese).

Dai, S.G., 1996. Environmental Chemistry. China Higher Education Press, Beijing, (in Chinese).

Fu, X.C., Shen, W.X., Yang, T.Y., 1990. Physical Chemistry, fourth ed. China Higher Education Press, Beijing, (in Chinese).

Gao, T.Y., Gu, G.W., 1999. Engineering for Water Pollution Control. China Higher Education Press, Beijing, (in Chinese).

Gu, T.R., Zhu, B.Y., Li, W.L., Ma, J.M., Dai, L.R., Cheng, H.M., 1994. Surface Chemistry. China Scientific Press, Beijing, (in Chinese).

Hiemenz, P.C., Rajagopalan, R., 1997. Principles of Colloid and Surface Chemistry. Taylor & Francis Group, New York, NY.

Ives, K.J., 1978. The Scientific Basis of Flocculation. Sijthoff and Noordhoff, Alphen aan den Rijn, The Netherlands.

Moore, W.J., 1972. Physical Chemistry. Longman Group Limited, London.

Rubin, A.J., 1978. Chemistry of Wastewater Technology. Ann Arbor Science Publishers Inc., MI.

ShawF D.J., 1978. Introduction to Colloid and Surface Chemistry, second ed. Butterworth & Co. (Publishers) Ltd., London.

Shen, Z., Zhao, Z.G., Kang, W.L., 2012. Colloid and Interface Chemistry, fourth ed. China Chemical Industry Press, Beijing, (in Chinese).

Stumm, W., Morgan, J.J., 1981. Aquatic Chemistry. John Wiley & Sons, Inc., New York, NY.

Tang, H.X., 1978. Chemical Basis of Water Supply and Drainage. China Architecture & Building Industry Press, Beijing, (in Chinese).

Yan, X.S., Fan, J.C., Xu, B.J., 1999. Engineering for Water Supply, fourth ed. China Architecture & Building Industry Press, Beijing, (in Chinese).

Zhou, Z.K., Gu, T.R., Ma, J.M., 1996. Basis of Colloid Chemistry. Beijing University Press, Beijing, (in Chinese).

Colloid and Interface Chemistry for Water Quality Control. http://dx.doi.org/10.1016/B978-0-12-809315-3.00015-3

Index

D

Q

R

S

Edwards Brothers Malloy
Ann Arbor MI. USA
July 25, 2016